IMAGES
of America

COOKE CITY

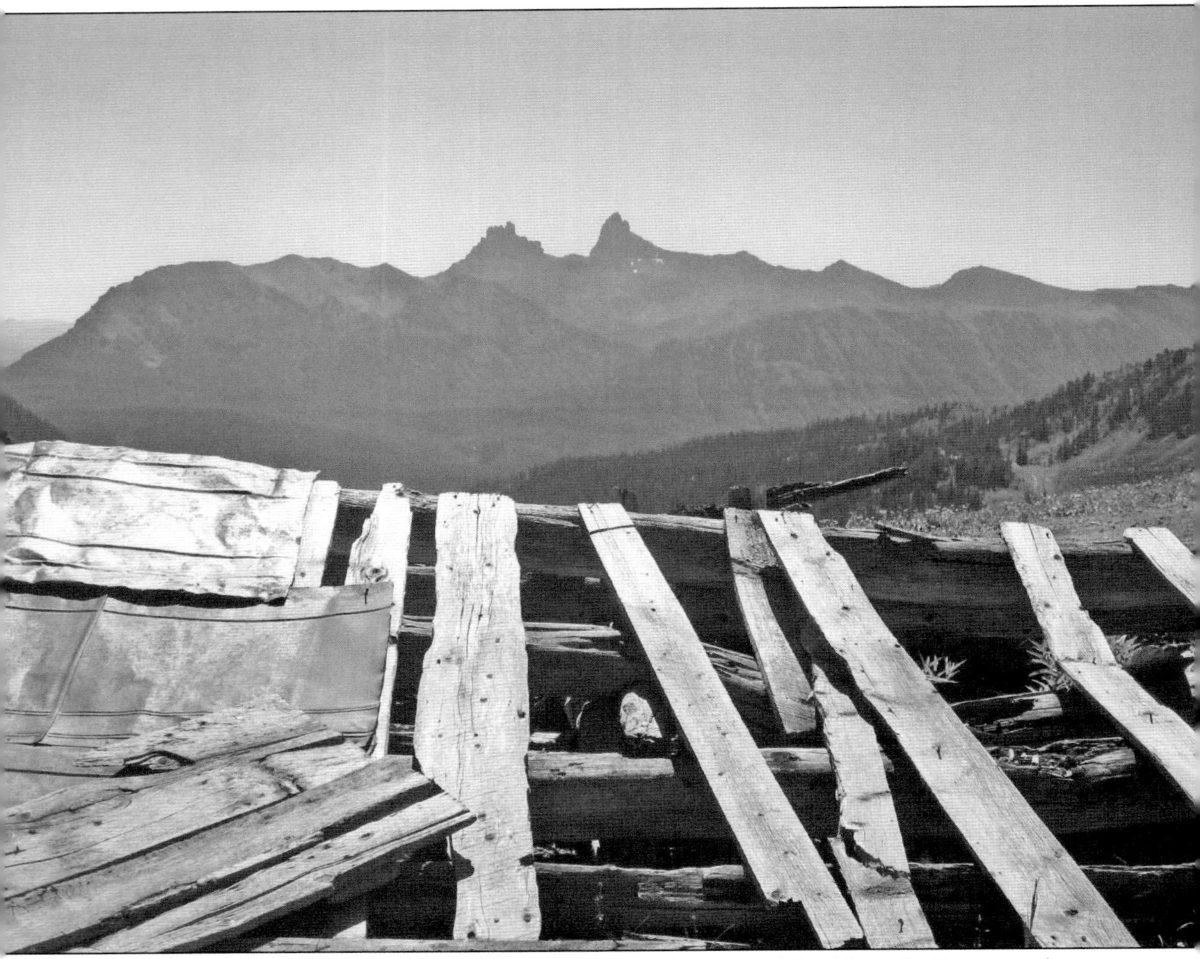

Index and Pilot stand as silent sentinels rising above the crags of the Absaroka Range to the west. Pilot, also known as the Castle, is a majestic mountain peak with granite turrets and spires surrounded by a granite fortress. All the gold and silver in the mountains cannot compare to their beauty. The world has greatly changed in their shadows. (Courtesy of Margaret I. Reeb Collection.)

On the Cover: Ingvold "Ing" Gronvold is standing on Henderson Mountain with Index behind him. Ing worked for Western Smelting & Power Company, operating the hydroelectric powerhouse in 1915. He attended three years of engineering school in Seattle and was in charge of installing the transmission lines. He hauled logs and ran the portable sawmill. In 1917, he left Cooke City to serve his country in the Navy, later returning. He is remembered for his collection of historic photographs and documentations and being instrumental in obtaining the photograph collection of the Dr. G.L. Tanzer family. (Courtesy of Margaret I. Reeb Collection.)

IMAGES
of America

COOKE CITY

Linda L. Holland

ARCADIA
PUBLISHING

ISBN 978-0-7385-7584-1

Published by Arcadia Publishing
Charleston, South Carolina

Printed in the United States of America

Library of Congress Control Number: 2011926553

For all general information, please contact Arcadia Publishing:
Telephone 843-853-2070
Fax 843-853-0044
E-mail sales@arcadiapublishing.com
For customer service and orders:
Toll-Free 1-888-313-2665

Visit us on the Internet at www.arcadiapublishing.com

This book is dedicated to the old-time miners for whom Margaret I. Reeb wrote a tribute on the cemetery sign:
"They lived and died for the dream these mountains gave them, and as in tribute . . . the wild forget-me-nots blossom every spring."

"Margaret's quote is not a mere romantic western legend. The beautiful blue forget-me-not flowers still bloom in the mountain valleys. It matters that this dainty flower regenerates each spring in a rugged high mountain altitude from under massive amounts of snow. . . . It is a symbol that reminds us of the human endurance, hope and the continued rebirth of nature and mankind."
—Dee Smith

Contents

ACKNOWLEDGMENTS

I am especially grateful to my editors, John Poultney, Elizabeth Bray, and Mike Litchfield, for their assistance and patience and to Arcadia Publishing for the opportunity to share some of Cooke City's history.

I am grateful to have had the opportunity to be a volunteer on the Whithorn Collection at the Gateway to the Yellowstone Museum. It gave me insight to appreciate the historic value of vintage photograph collections. Thanks to director Paul Shea and Laura McCarthy of that museum, Gallatin Historical Society, the Carbon County Historical Society, Gallatin National Forest and Jane Ruckman, and National Park Archives. I am thankful for the Park County historians. And to Dee Smith, your enthusiasm of Cooke City history, knowledge, and witty sense of humor energizes me. Thanks to Brenda for her support and skills. Most importantly, Randy, I thank you for your patience, support, and a quiet space. I take personal responsibility for any errors in this publication.

The photographs in this publication are from the Margaret I. Reeb Collection unless otherwise noted. May Margaret be recognized for her careful preservation of extensive and marvelous historical documentation for future generations.

INTRODUCTION

The area around Cooke City was inhabited by Native Americans, and later came the explorers and fur trappers. In 1803, the United States bought the Louisiana Purchase from France, and the historic expedition of Lewis and Clark explored the new territory. But by the stories of geysers, hot springs, and a map, some believe John Colter was the first white man to actually to explore the Yellowstone area. In 1864, a party of men with George A. Houston traveled up the Yellowstone, and a part of that group reached the head of the Soda Butte Creek before being attacked by Indians. In the fall of 1869, Adam "Horn" Miller organized a prospecting party into the area, and the era of prospecting and mining began.

On the Crow Indian Reservation, an illegal furnace of the Mexican type was built in 1875, and lead ore from Miller Mountain had been smelted. Still in the furnace, lead bullion was stolen by a raiding party of Nez Perce, presumably to be converted into shot. On their way to Canada, Chief Joseph and his band were being pursued by government troops. This furnace was marginal.

On April 11, 1882, some of the reservation was relinquished, and the rush was on. Cooke City was the typical rough-and-tough mining camp, with saloons, gambling, fighting, murders, prostitution, drunkenness, wife beating, and claim jumping. Life had its hopes, struggles, despair, and chicanery.

Cooke City naturally had what was necessary for mining: plenty of water and timber. In addition, it had valuable beds of tile-and-brick clay, limestone, and building stones nearby. Transportation had always been a problem, being isolated high in the mountains with snow cover many months of the year. The district was considered one of its most promising camps in Montana, but shipping costs were prohibitive.

A transcontinental railroad financed by Jay Cooke inched its way across the country. The mineral deposits in the area attracted eastern interests, including that of Cooke, who vowed to build a railroad to the mining camp named after him. In 1873, Cooke was blamed for the worst financial panic since 1837, and the Northern Pacific Railroad went into receivership. Cooke's financial disaster and the dedication of the first national park put an end to the easy railroad route to the New World Mining District. The Park Branch of the railroad was constructed to Cinnabar and later to Gardiner at the northern entrance of the park. Cooke City attempted to get a railroad for 40 years.

In 1902, there was a stir in Cooke City when the government improved the old road from Gardiner through Yellowstone National Park to Cooke City.

Mining activity waxed and waned for many years. The great mines at Butte came into prominence and attracted big capital. Over the years, the speculative world heard of Cooke City. The gullible public all over the world was mined by businessmen selling Cooke City stock.

Heap roasting, roasting furnaces, stamping, and cyanide mills were short lived. The Western Smelting & Power Company with its modern technology brought hope to Cooke City mining. The Homestake, Daisy, and Alice E. Mines provided the majority of the minerals produced. The

Glengarry Mining Company engaged in developing some copper and gold. Miners came and went, and mining companies changed owners and names. Few found the wealth they searched for.

However, in the 1920s, the New World Mines persevered to become Montana's third most productive district. Walter McClaren's open-pit mine was one of the most prosperous gold mines. This mine produced until the 1950s. Subsequently, drilling by Crown Butte in the early 1990s identified that the exploration by Western Smelting & Power was near the "mother lode." The dreams of the rugged early individuals who pioneered the Cooke City country were finally realized.

The following pages of pictorial history of the early years will give you an insight of the struggles of mining and the community of Cooke City. Mining left its scars on the area. If Cooke City had gotten a railroad, if the price of minerals had stayed high, if operational costs had not been too expensive and the minerals were endless, what would the area be like today?

Along with the slow decline of Cooke City's mining tradition came the introduction of the automobile into Yellowstone National Park. Better roads were built, and the true wealth of the area was revealed in the scenic wonders and the rich character of the people who live there.

One

In the Beginning

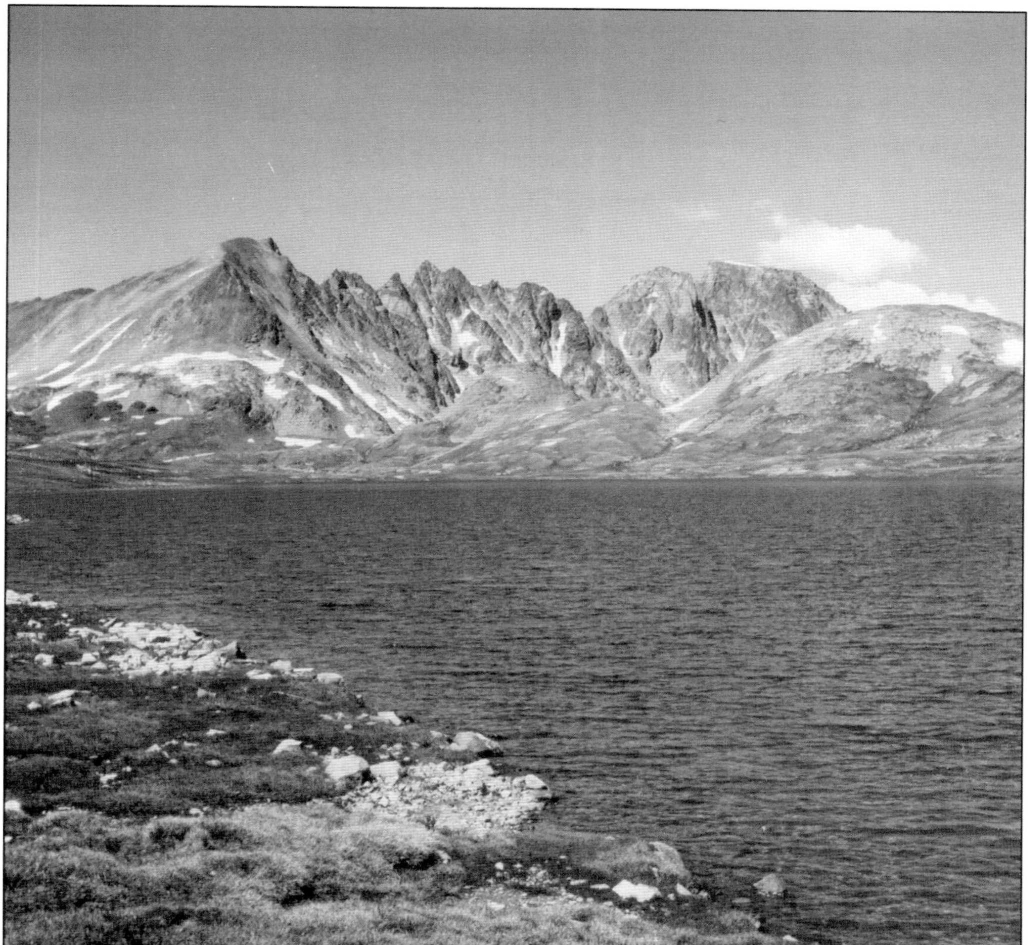

Goose Lake is located in a glacial basin surrounded by the Beartooth Mountains. The Crow Indians gave the name to the most prominent peaks because of their resemblance to vast molar teeth. The most prominent peak in the range is Granite Peak. At an elevation of 12,799 feet, it is the highest mountain in Montana.

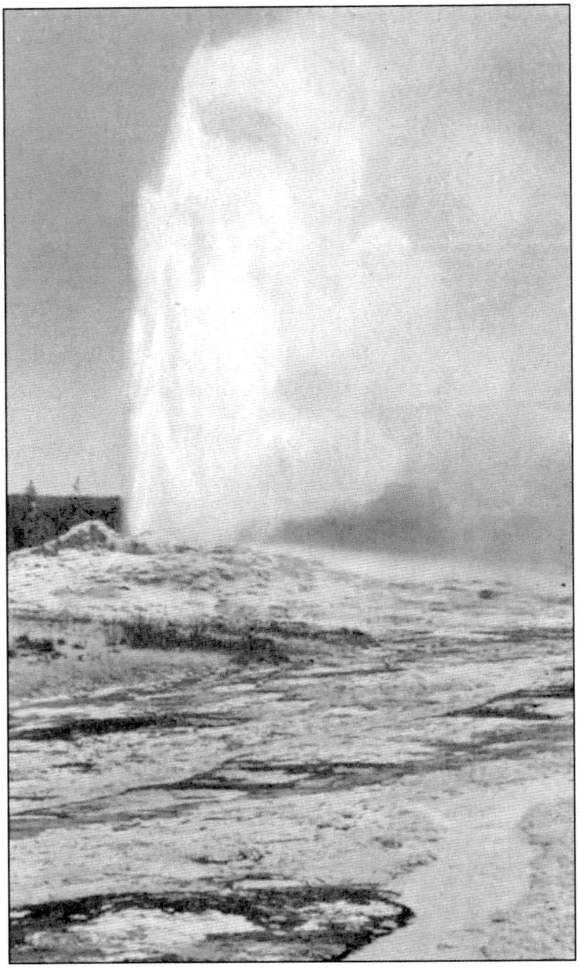

Reports of the hydrothermal wonders of Yellowstone created interest, and a series of expeditions were made resulting in the preservation of Yellowstone National Park as the world's first national park in 1872. Old Faithful geyser is one of the most recognized of "Wonderland's" geysers.

In this vintage photograph are terraces at Mammoth in Yellowstone National Park, another thermal wonder. Hot water rises through limestone deposits, sculpting these travertine terraces.

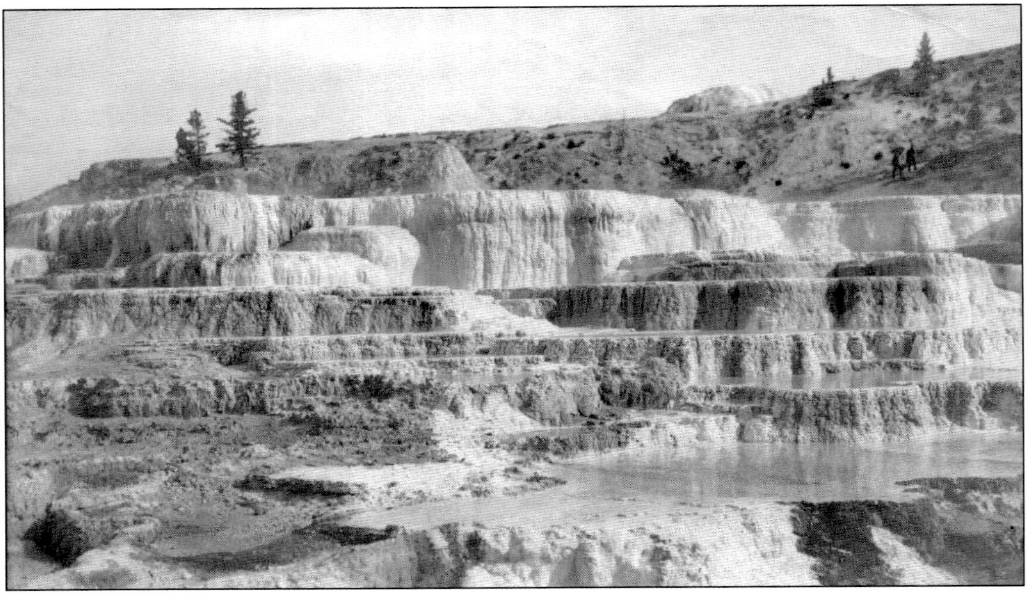

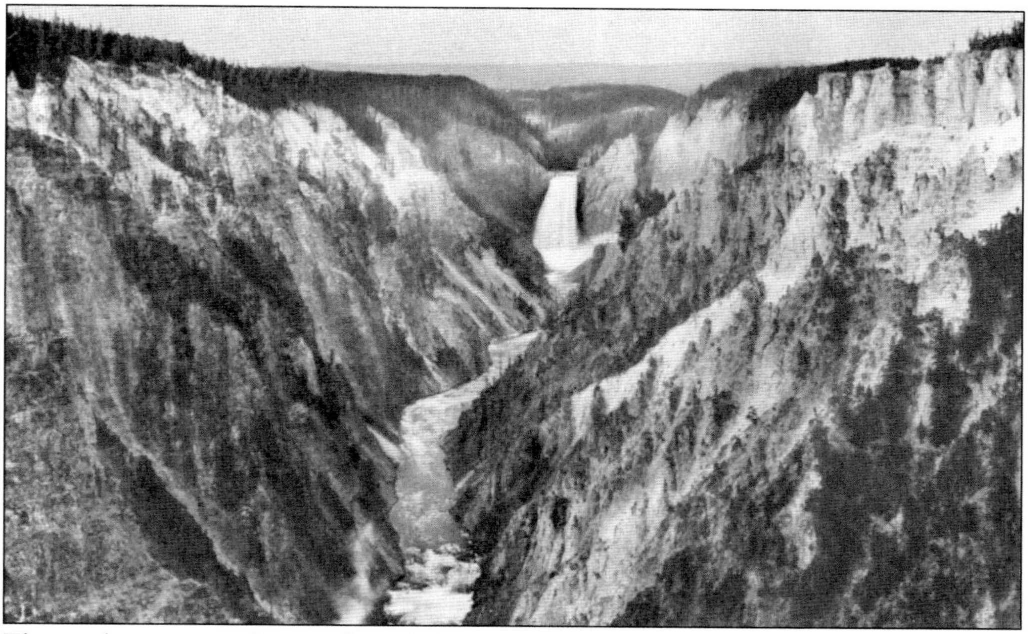
The eroding power of water shaped the Grand Canyon of the Yellowstone, exposing its rainbow colors.

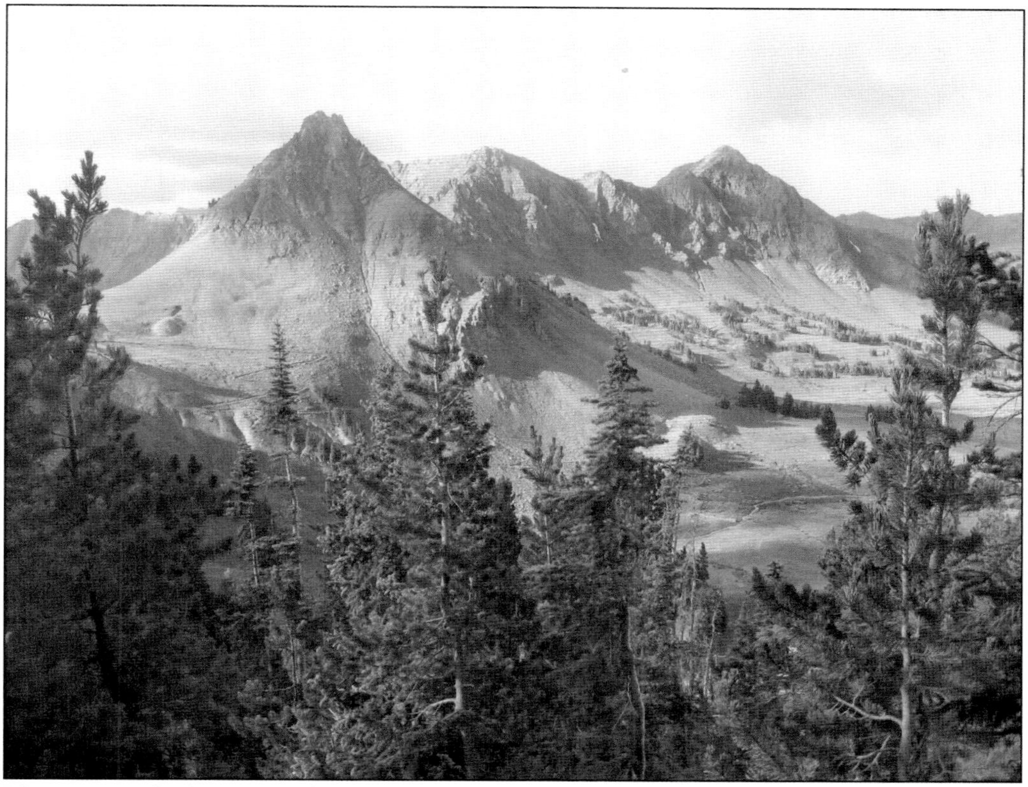
This image is looking toward Daisy Pass above Lake Abundance.

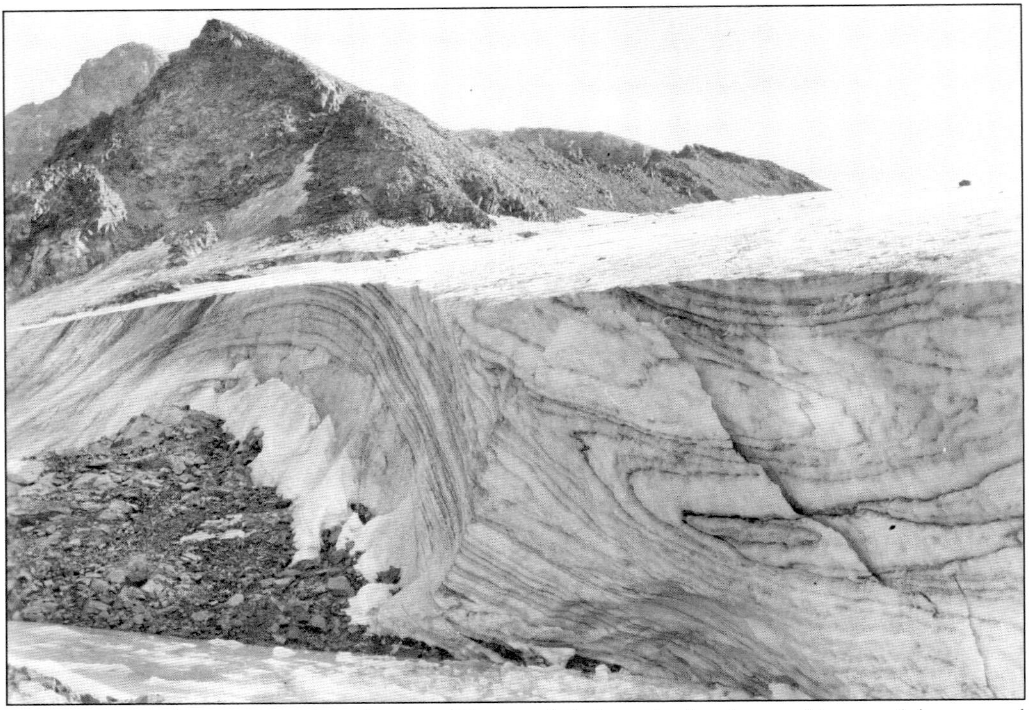

Found in rugged mountains, Grasshopper Glacier is left from the Ice Age and is one of the natural wonders of the world. Grasshopper remains are embedded in the ice. It is believed that high-flying grasshoppers were brought down by one of the area's snowstorms. Grasshopper Glacier is one mile northeast of Goose Lake at the head of Rosebud Creek. (Courtesy of Gateway to the Yellowstone Museum.)

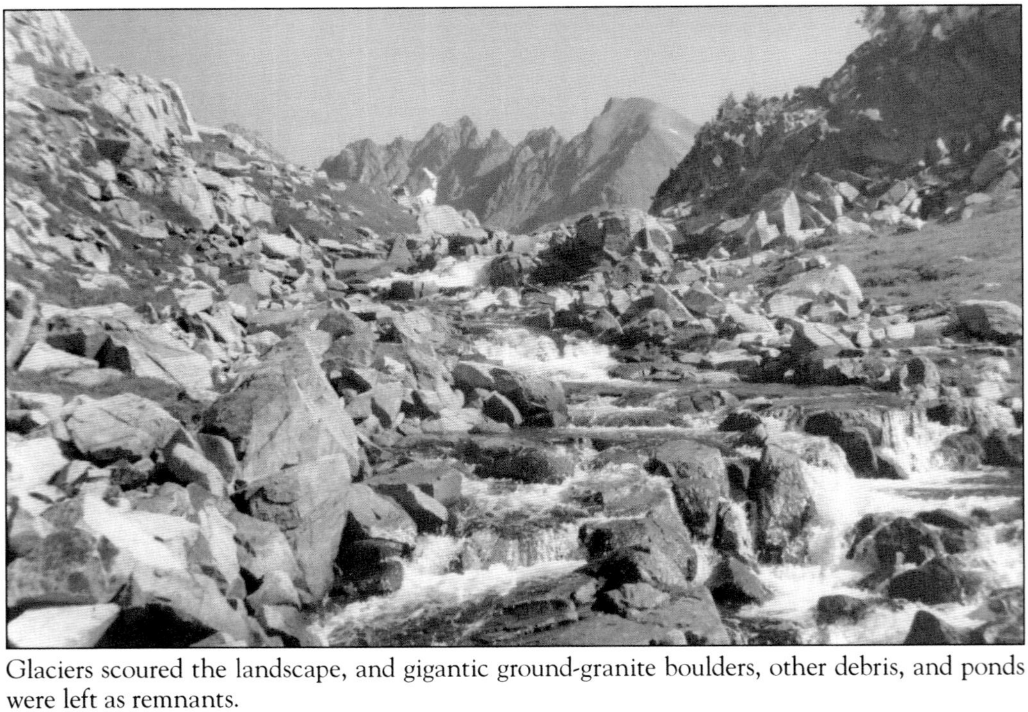

Glaciers scoured the landscape, and gigantic ground-granite boulders, other debris, and ponds were left as remnants.

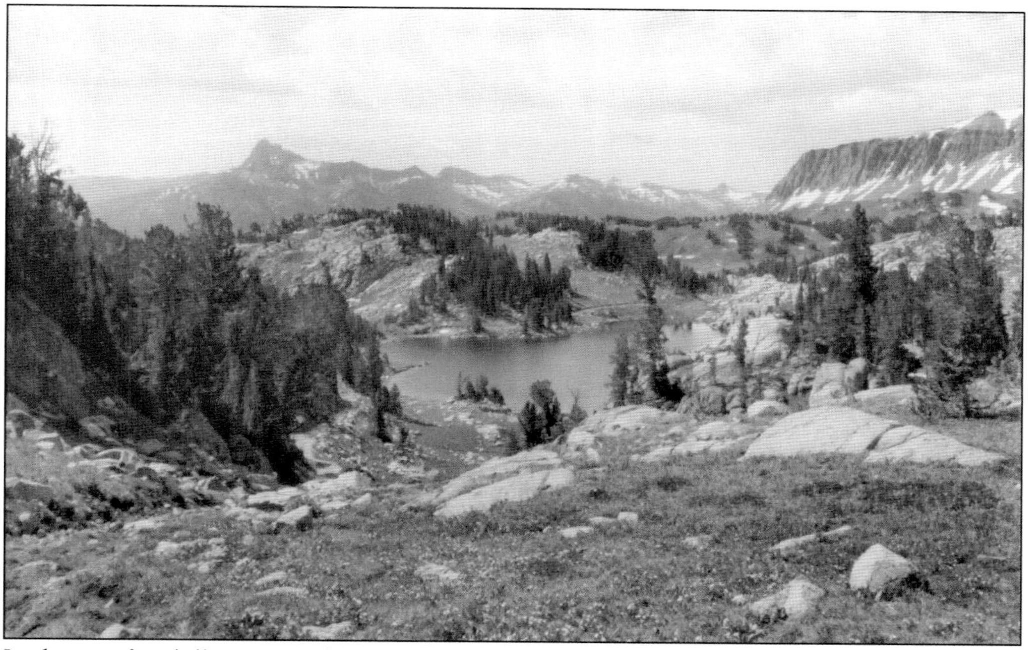

In the past four billion years, the mountains were formed by tremendous pressure and heat. The earth cooled. In the past 75 million years, compressing, folding, and uplifting occurred. Volcanoes spewed, mountains ranges were uplifted and fell, ice, wind, and water shaped lakes, and streams and waterfalls were created. The forces of creation are evident in this photograph.

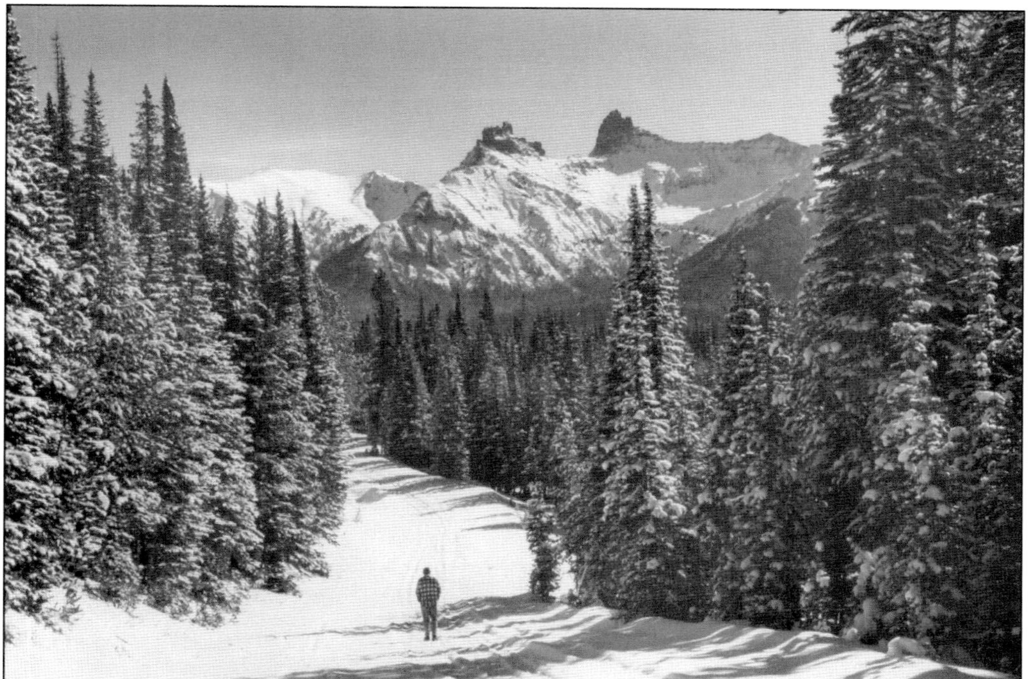

When A. Bart Henderson came to the area with "Horn" Miller and prospector friends, he recorded the names of Pilot and Index in his journal. They can be seen from a great distance and were used by prospectors, the military, and pioneers as a reference point.

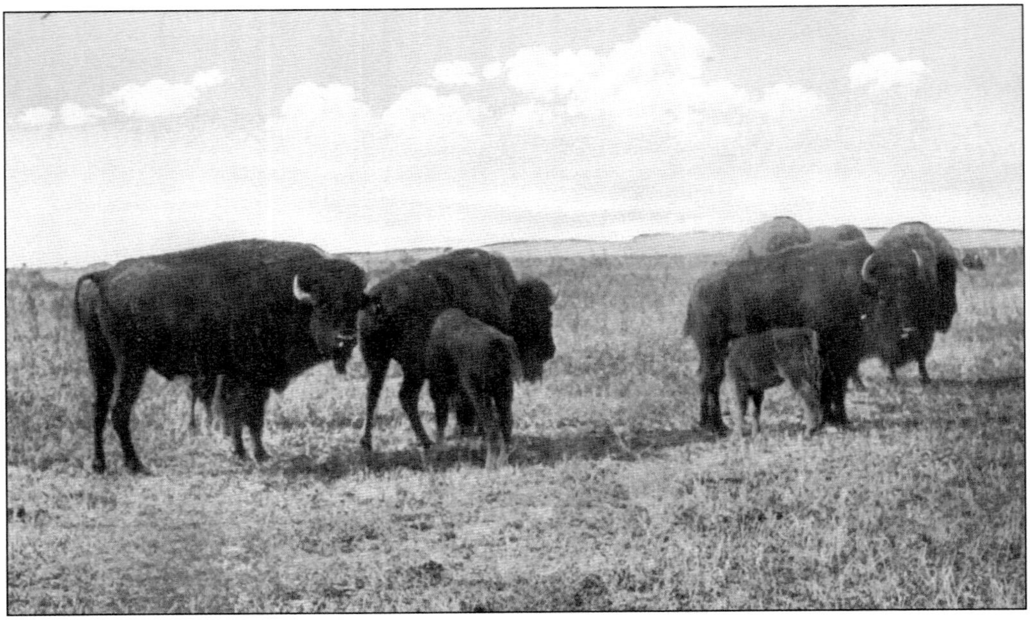

Wildlife in the area is abundant. It includes buffalo, black bear, grizzly, moose, mule deer, big horn sheep, mountain goats, coyotes, and the reintroduced wolf. In the postcard above is the buffalo, the beast that once roamed the West. The grizzly, shown in the vintage postcard below, is the most feared of the large mammals.

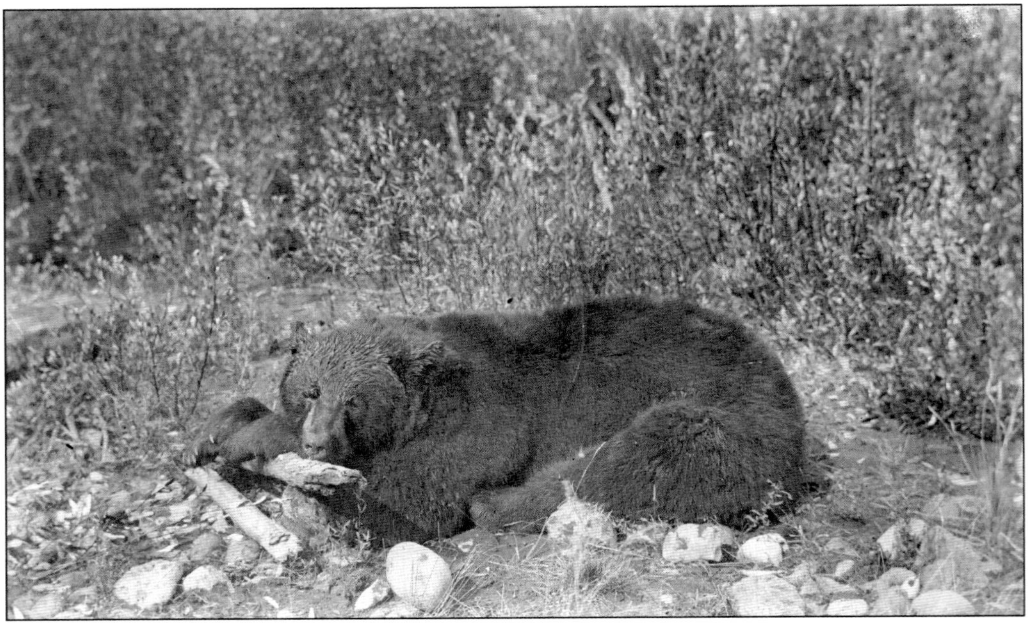

Two

GALENA

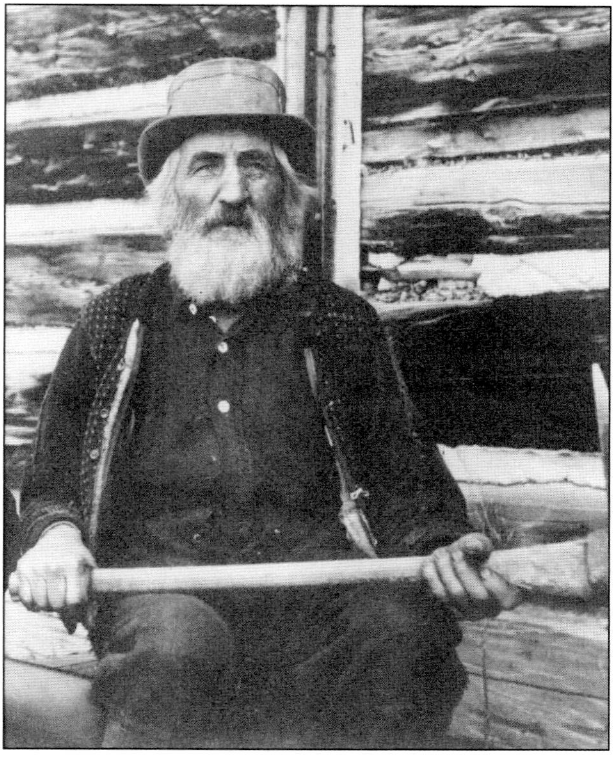

Adam "Horn" Miller was a trapper, buffalo hunter, and prospector as well as an Indian fighter. The first reliable record of the discovery of gold was made by Miller and prospecting friends J.H. "Pike" Moore, James Gurley, and Bart Henderson. In the fall of 1869, Jack Crandall had shown ore samples to his friend Horn Miller in Bozeman. In the spring of 1870, Miller and partners were robbed of their horses by Indians. They cached what they could not carry on what today is known as Cache Creek and narrowly escaped death. That year, Horn returned with a group of prospectors and discovered gold in the Clark's Fork and Soda Butte drainages and staked the first two claims in the area—one was the Shoo-Fly. The men named the area the New World Mining District and were the founding fathers of Cooke City.

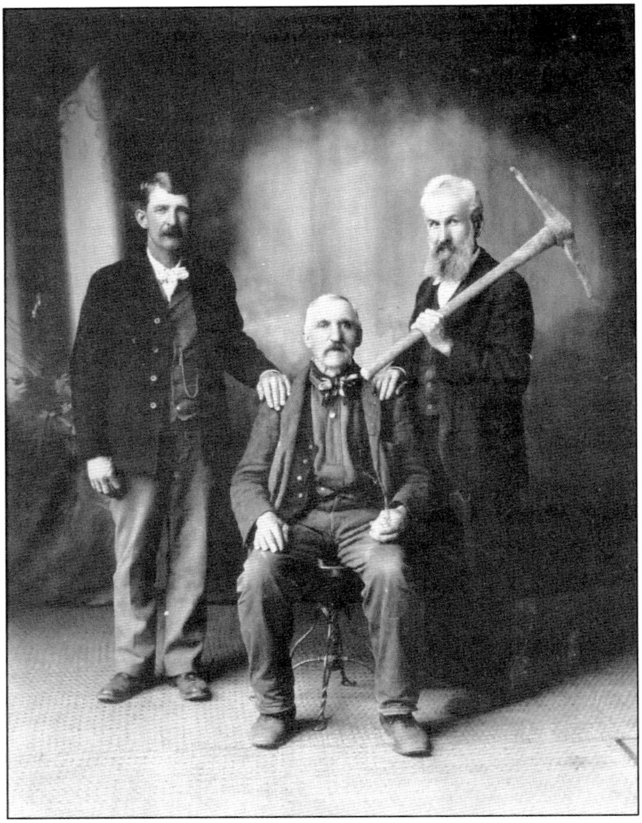

Pictured here from left to right are John Curl, Adam "Horn" Miller, and Joe Brown. They helped Bart Henderson build a wagon road through a canyon along the Yellowstone River. Henderson recognized the economic potential of hauling mining supplies to Cooke City and ore down the Yellowstone. (Courtesy of the Yellowstone Gateway Museum.)

"Yankee Jim" George seized the wagon road, built a way station, and collected a toll from travelers. Yankee Jim's name remains on that canyon today. The creek across the Yellowstone from Yankee Jim's was named for Joe Brown. Names of mountains and creeks may be the only monuments to the memory of individuals who had an impact on the West. (Courtesy of the Yellowstone Gateway Museum.)

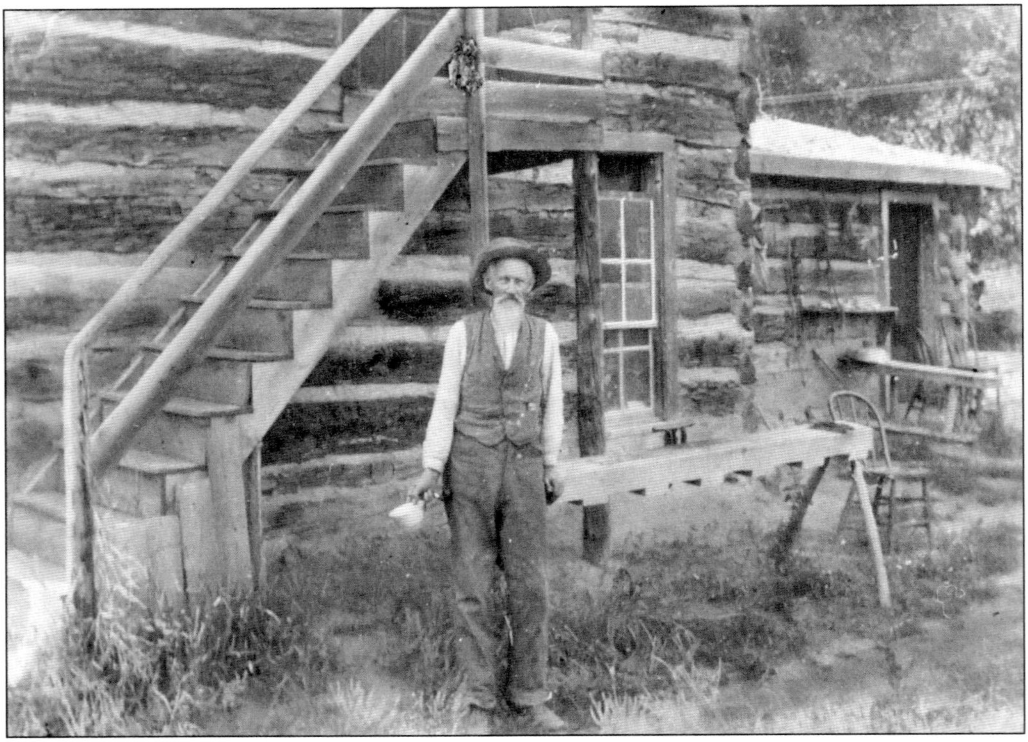

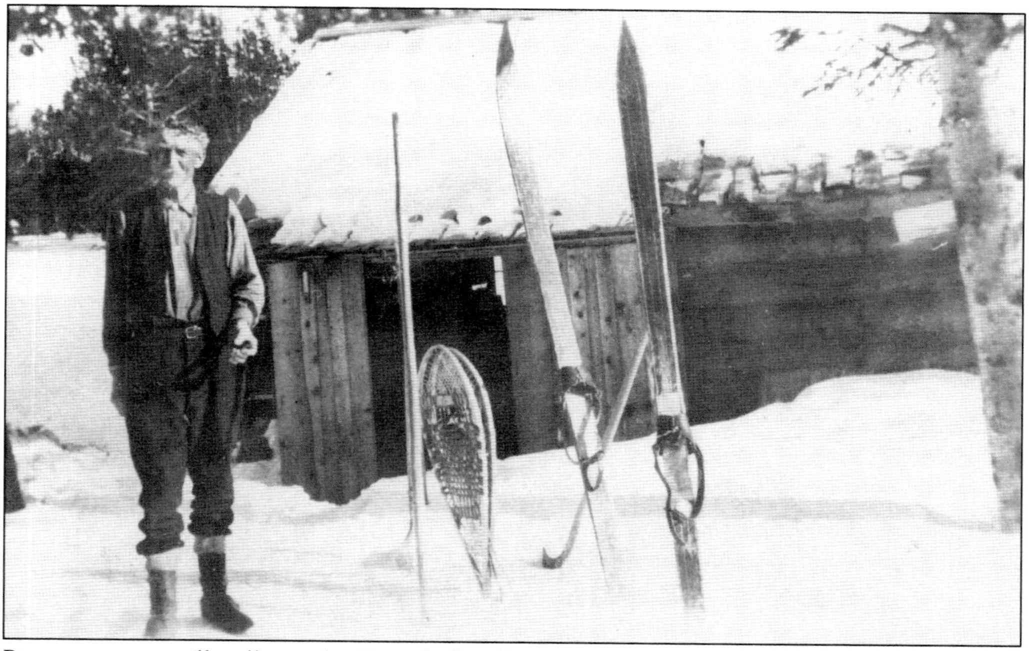

Prospectors were illegally on the Crow Indian Reservation prior to 1882. When the boundaries were moved, area was left to the public domain. The news of the wealth of the area brought more prospectors racing in during the winter. They came by horseback, foot, and skis, and others were challenged by Norwegian snowshoes. Nearly 1,500 claims were made and recorded in the district, but many were allowed to lapse.

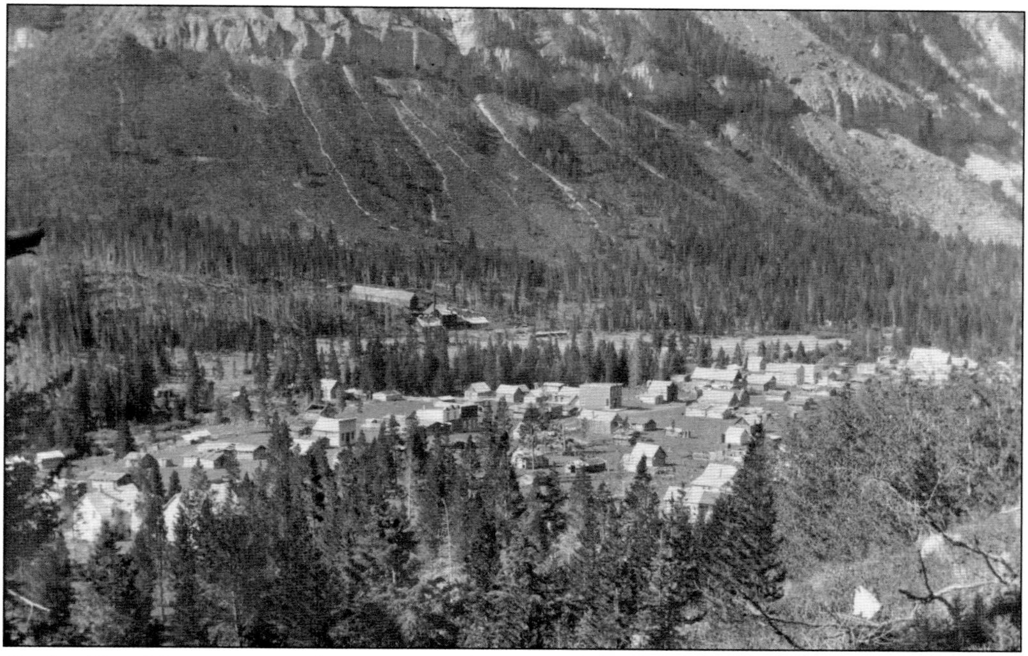

Cooke City, Montana, was the commercial center of the Cooke Mining Camp and Republic Mountain. On June 11, 1880, the Town Site Company took charge of the town's property and had it platted by 1883. There was indiscriminate "jumping of town lots," and there were no clear titles to the lots until 1890.

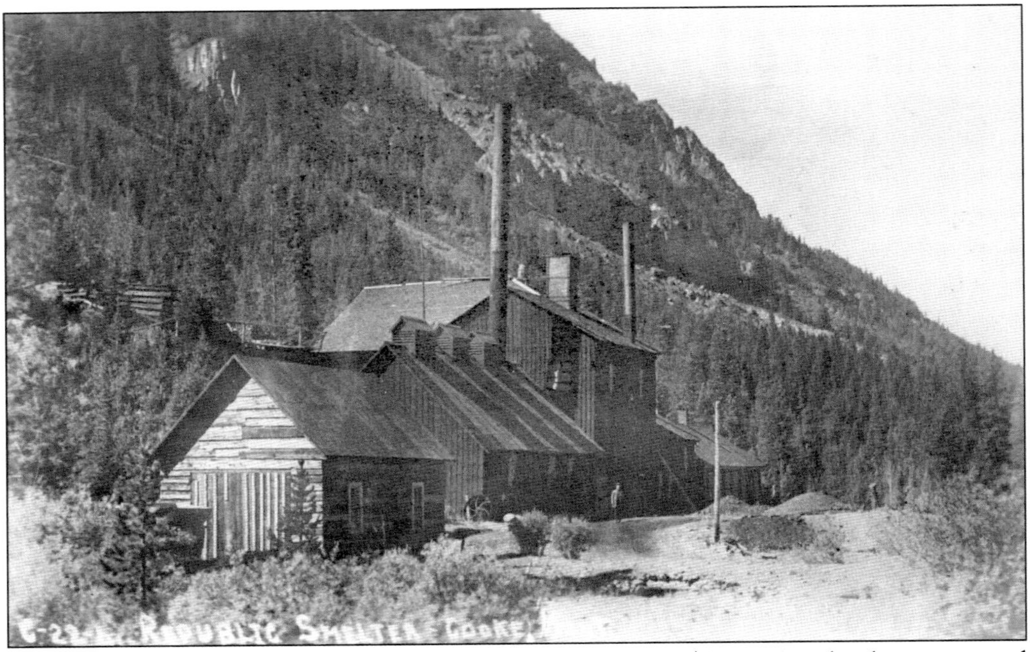

Maj. George Eaton formed the Republic Mine in 1883, investing $300,000 to develop a mine and smelter. After setbacks, the Republic was successful in producing tons of silver-lead bullion in 1885. The profits were cancelled by high freight charges. In 1886, mining activity slowed down, the principle ledges were blocked, and ore was put in mine dumps.

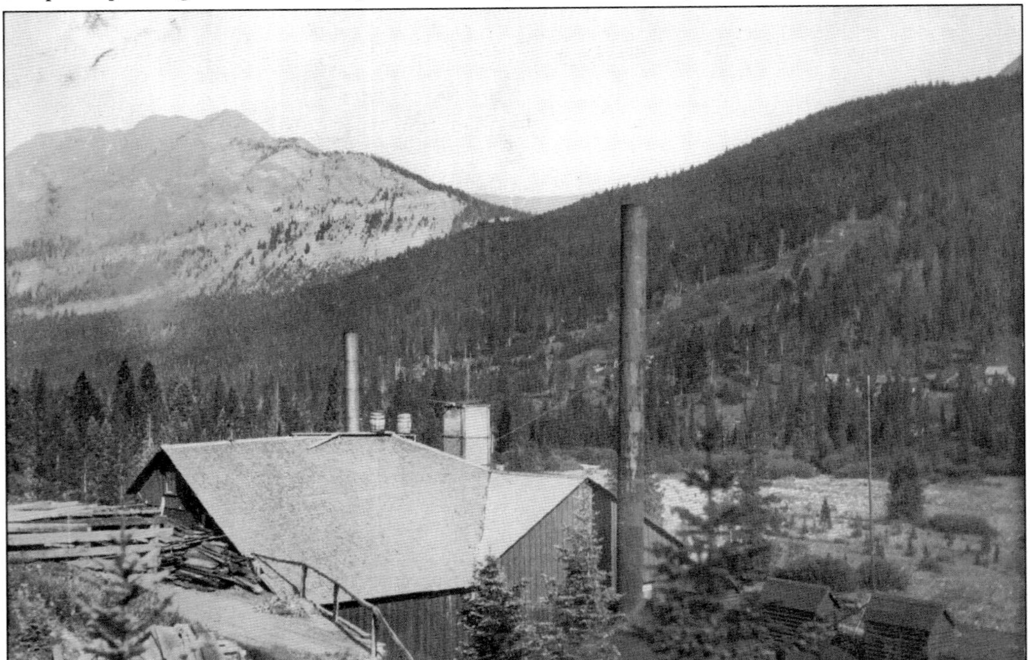

This image looking north toward Miller Mountain shows the Republic smelter. Without coal available, charcoal was made by burning stacked piles of green timber covered with wet branches. It was allowed to smolder for days until the timber had been consumed and wood charcoal remained. Charcoal would not burn hot enough for some processes and was later outlawed.

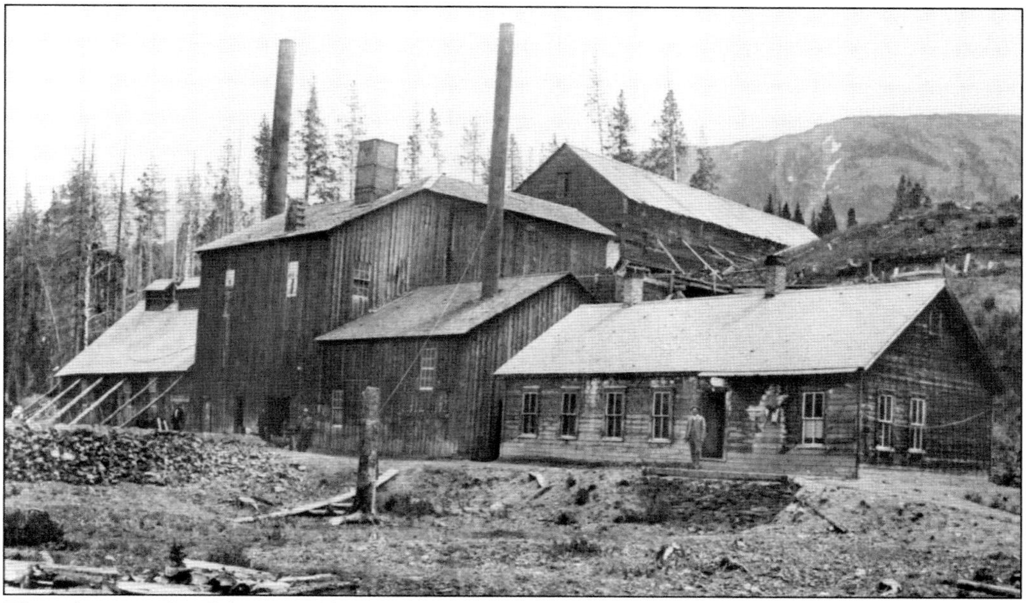

This photograph of the Republic smelter was taken about 1906. The Buffalo Montana Mining and Milling Company bought the Republic Mine, owned by R.I. McKay. (Courtesy of Yellowstone Gateway Museum.)

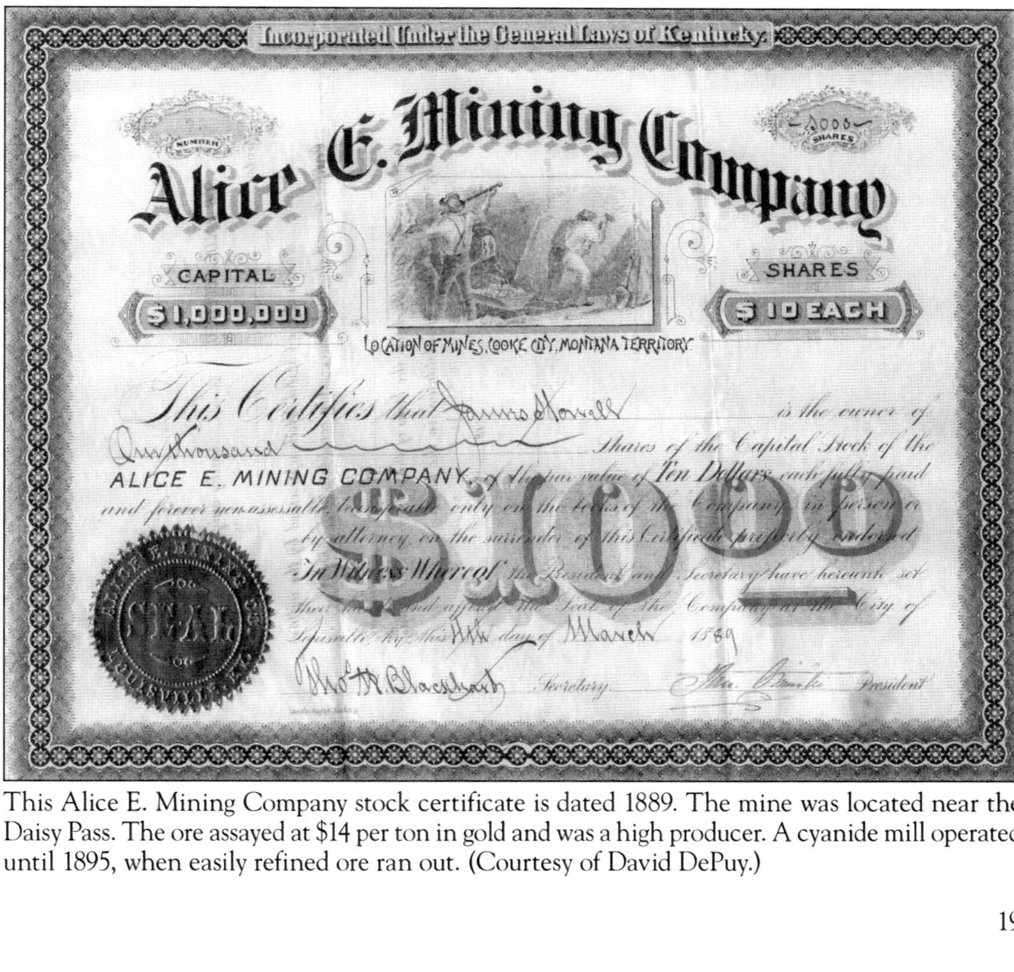

This Alice E. Mining Company stock certificate is dated 1889. The mine was located near the Daisy Pass. The ore assayed at $14 per ton in gold and was a high producer. A cyanide mill operated until 1895, when easily refined ore ran out. (Courtesy of David DePuy.)

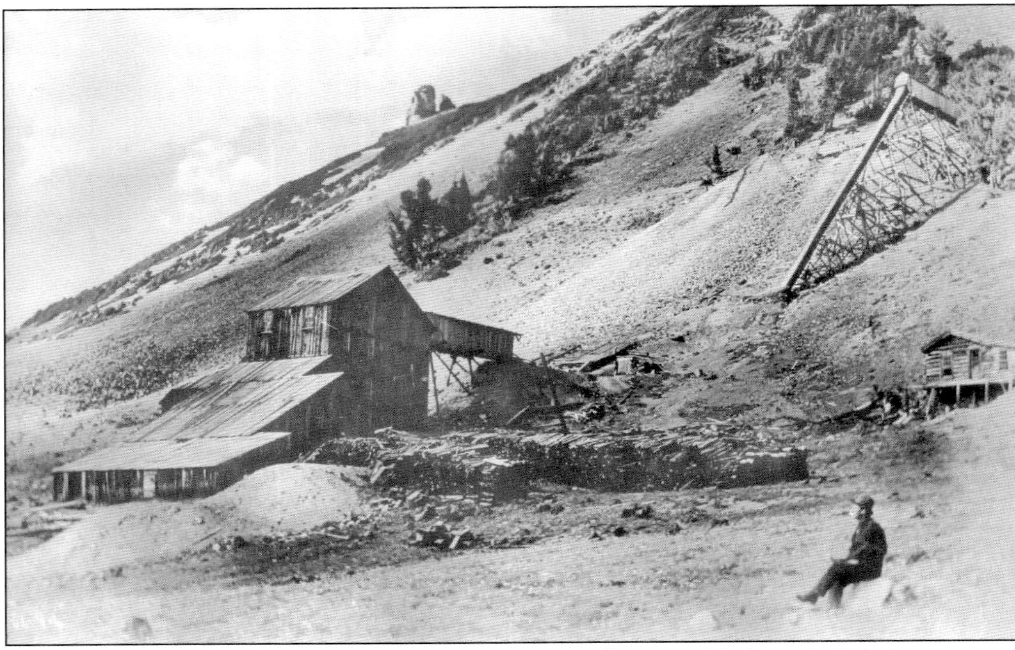

Miners grew gray and stooped, but many never lost hope and believed, "Maybe just over the next hill."

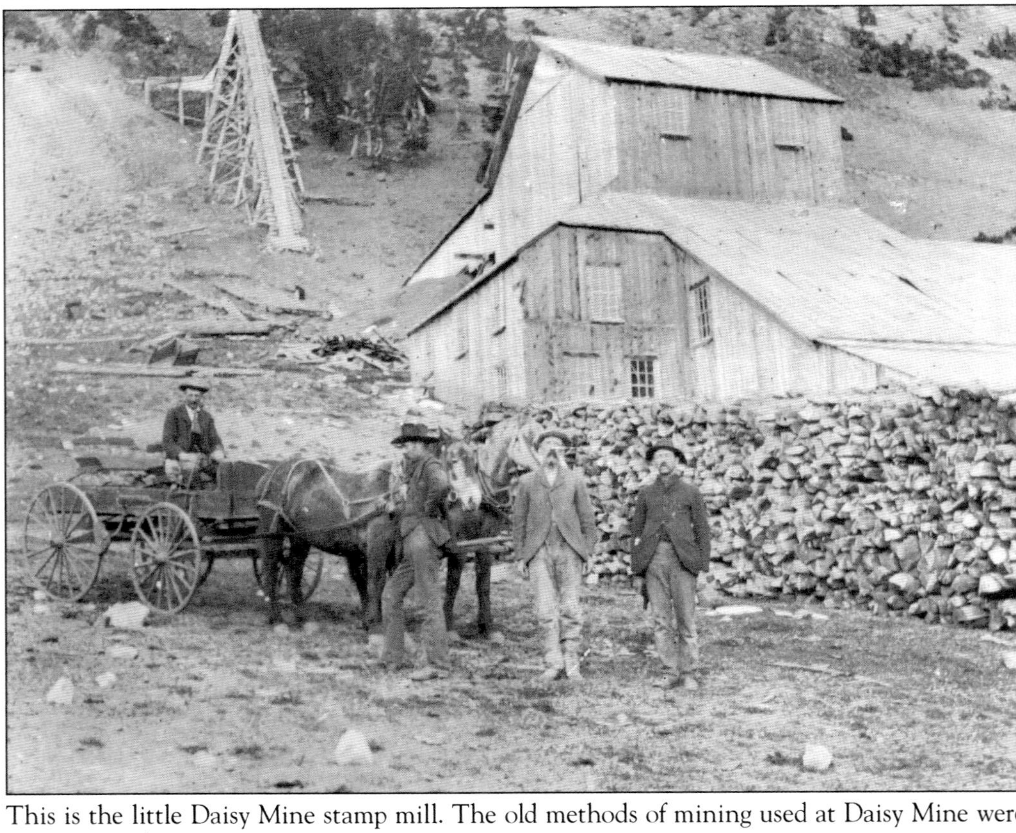

This is the little Daisy Mine stamp mill. The old methods of mining used at Daisy Mine were not very productive.

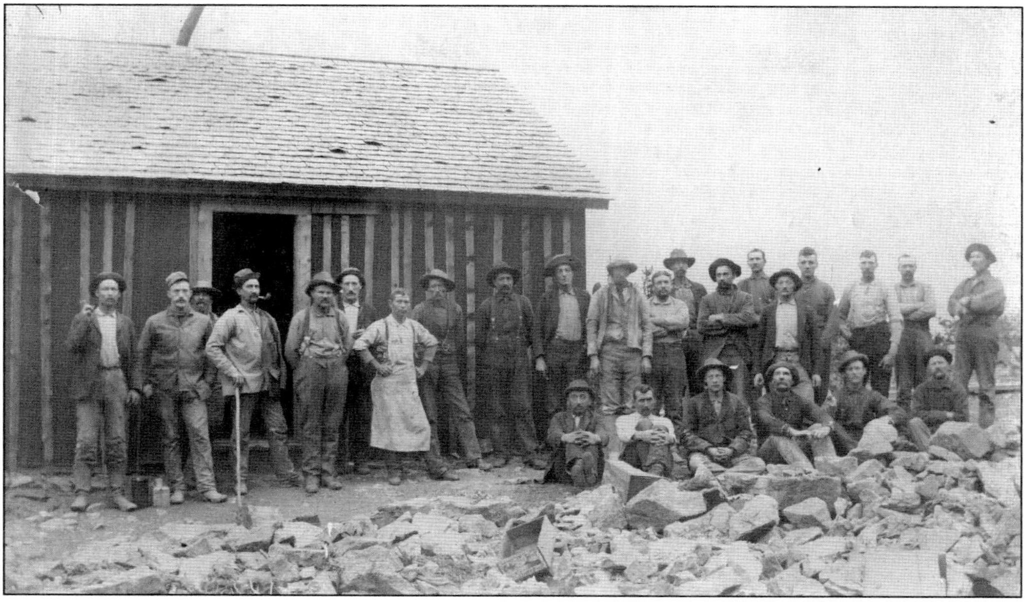

"Daisy miners during the 80's" is written on this photograph.

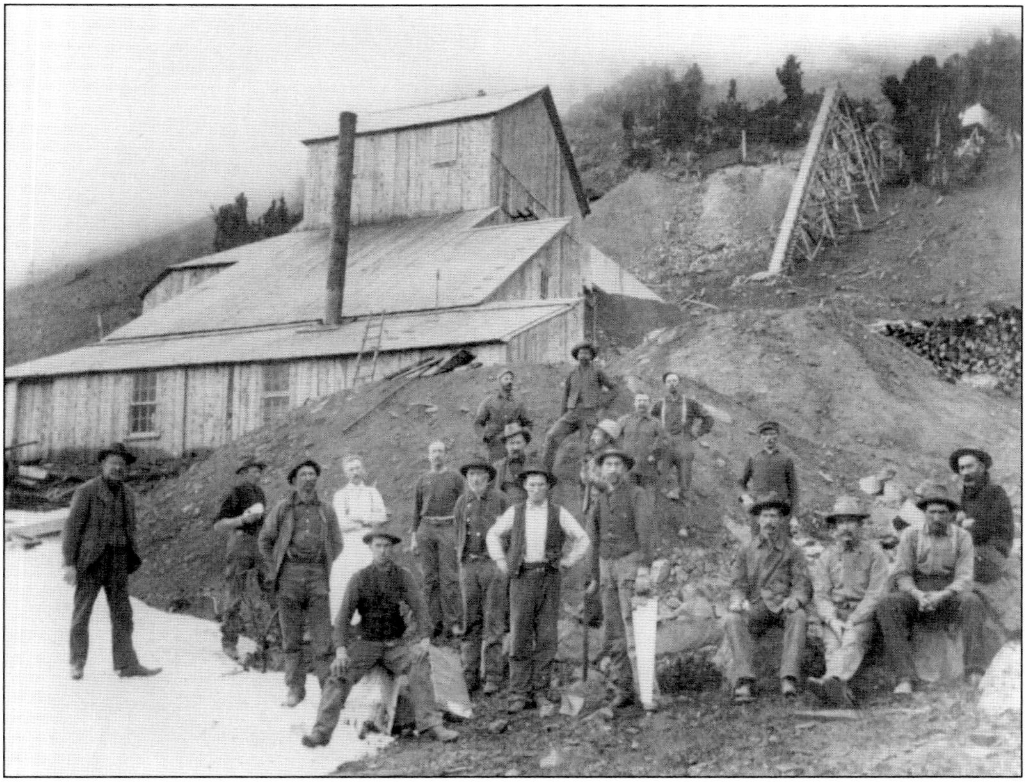

Daisy's first ore was shipped to Omaha, and a small 10-stamp mill was constructed in the early 1890s. When silver prices dropped, causing a national depression, the mine and mill ceased production. The Daisy stamp mill was in front of the old two-story Blackhart house on Broadway but is now an exhibit at the Cooke City, Montana, Museum.

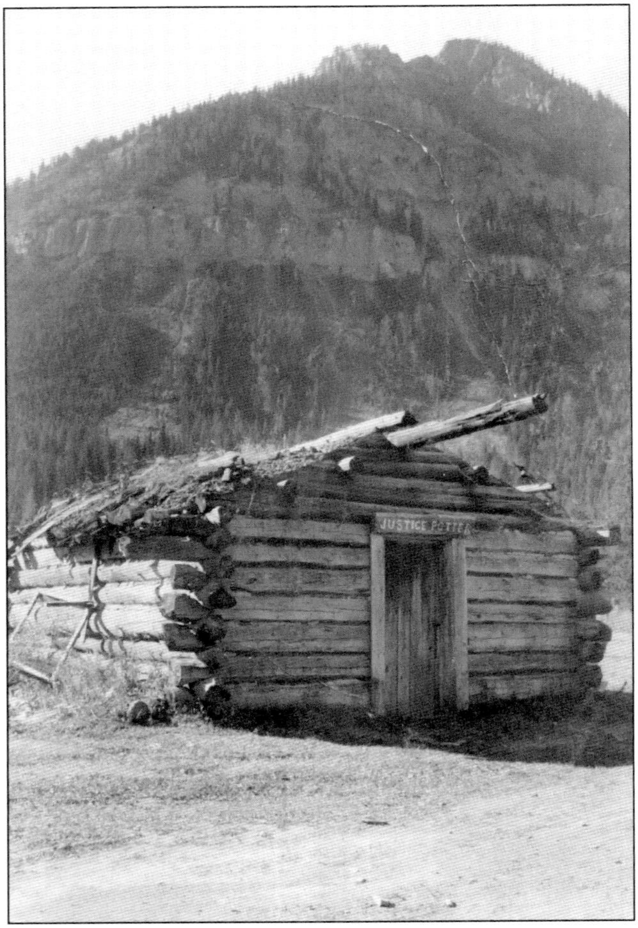

The justice of the peace was one of the first business houses. Certificates of location of the Lode Mining Claims had to be signed by the justice of the peace. Affidations of representation of Quartz Lode also had to be notarized. These proclamations verified that labor and improvements were made to the claim.

The sign on this early photograph identifies the Allen Hotel, known later as the Cosmopolitan. A grand piano was freighted in from Bozeman, and the ivory keys tinkled out familiar tunes.

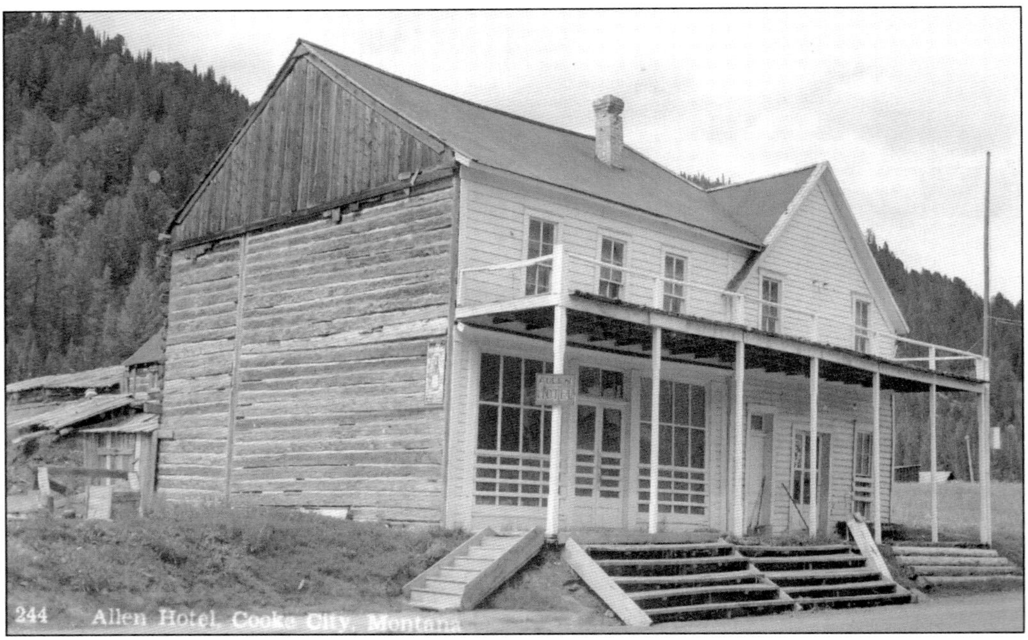

Mount Republic rises high above this rustic cabin built along Soda Butte Creek. In the 1880s, Cooke City had about 135 cabins, tents, 2 general stores, and 13 saloons.

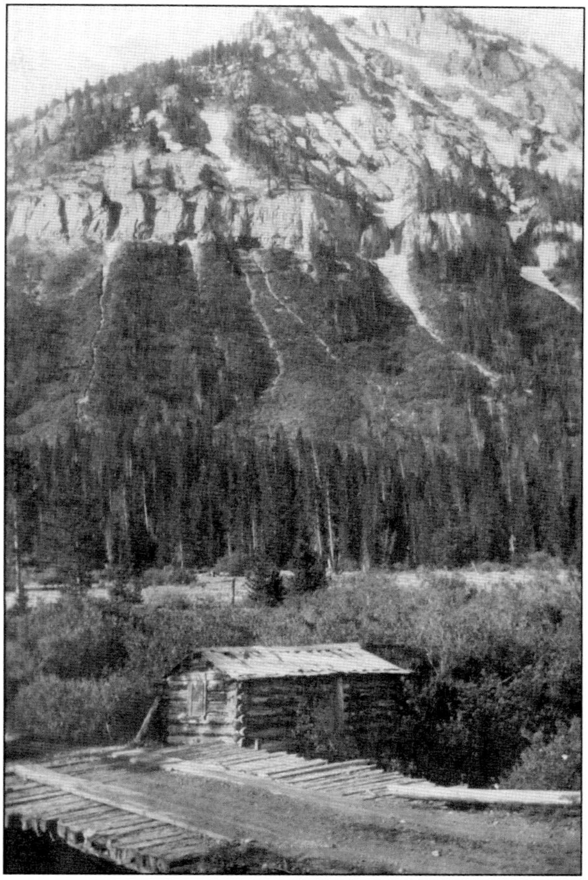

The Cosmopolitan Hotel was the social center of town; men gathered in the office to examine their specimens of ore and dream of wealth.

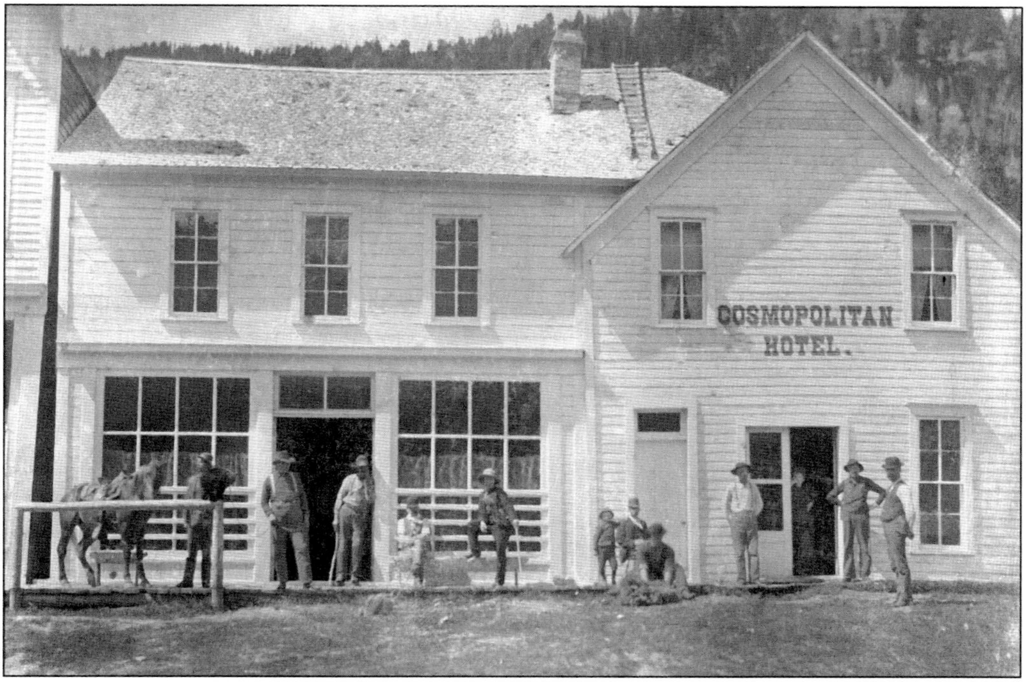

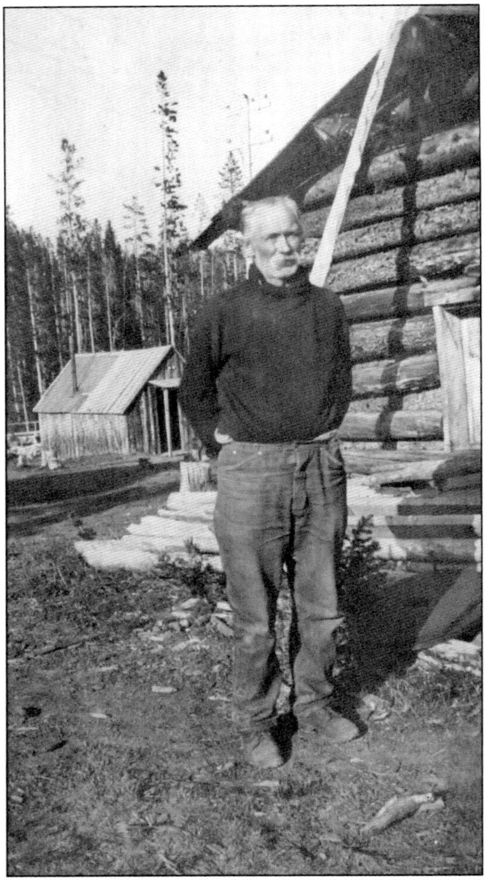

William Nichols went to early Cooke City with the hopes of striking it rich and seldom left in 40 years. Once, the judge called him to Livingston briefly for district court. Seeing how the outside world had changed, he quickly returned to his log cabin. He later sold his mine, but with Prohibition, it has been said he found little to spend his money on.

Horn Miller (right) poses with prospector friends.

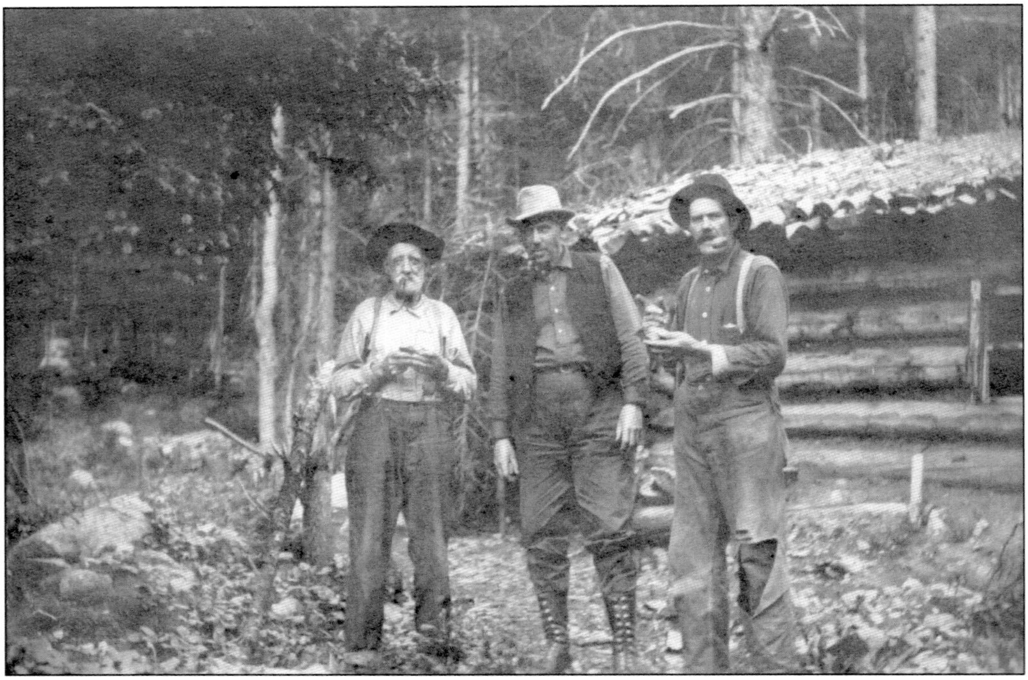

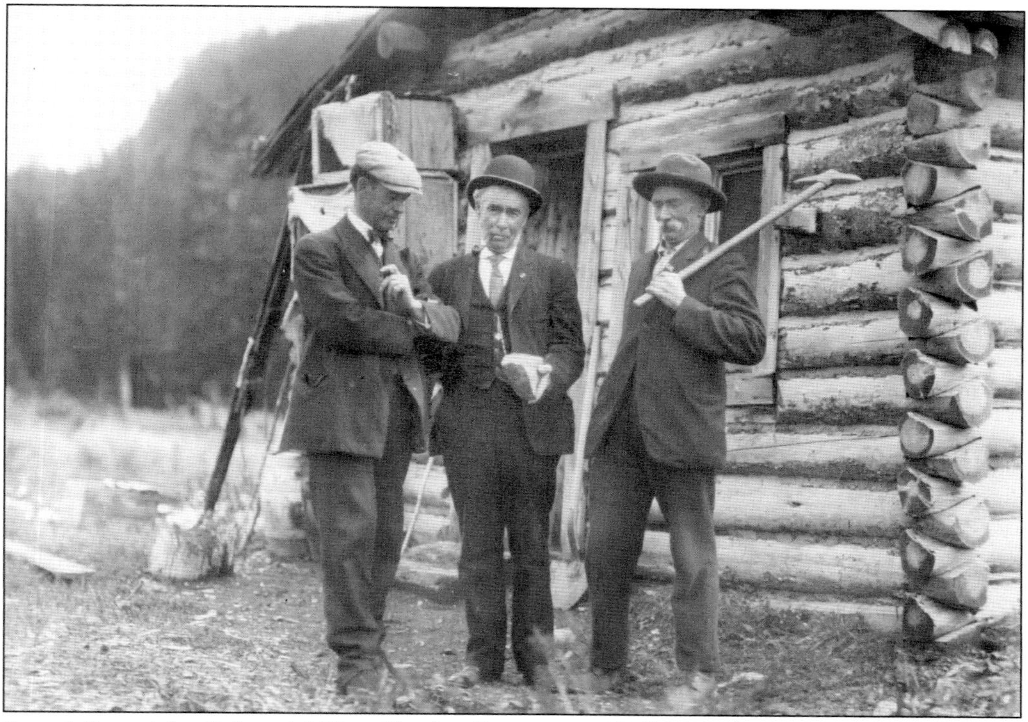

From left to right, Robert McKay, Charles Cannon, and Bill Nichols inspect a piece of quartz in the 1880s.

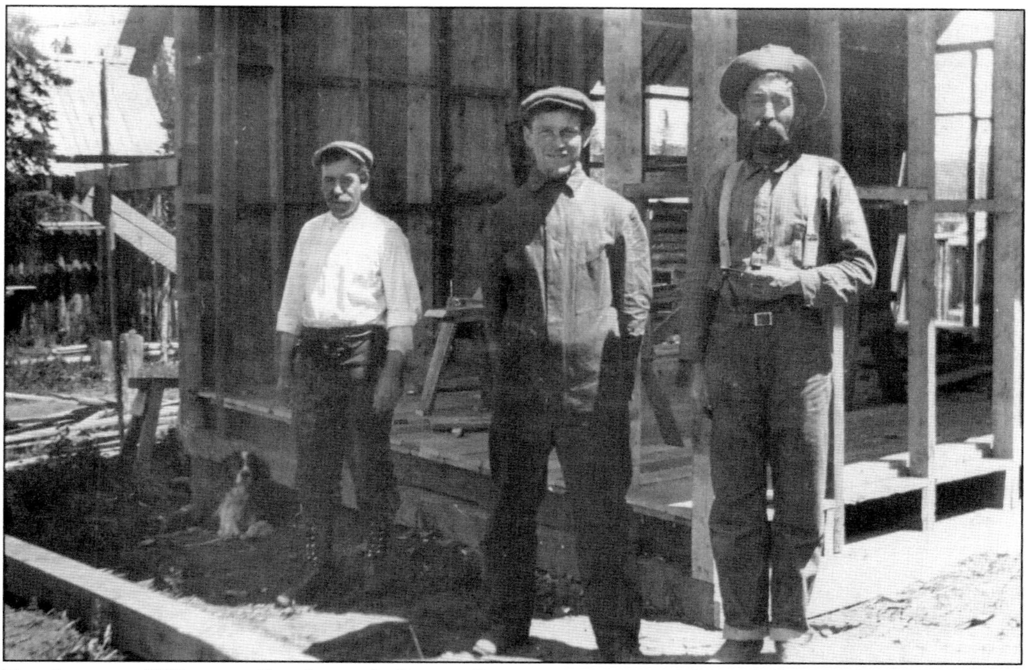

Employees of Western Smelting & Power Company, pictured from left to right, are sawmill superintendent H.C. Reagan, Bill Woelke, and Jim Cole.

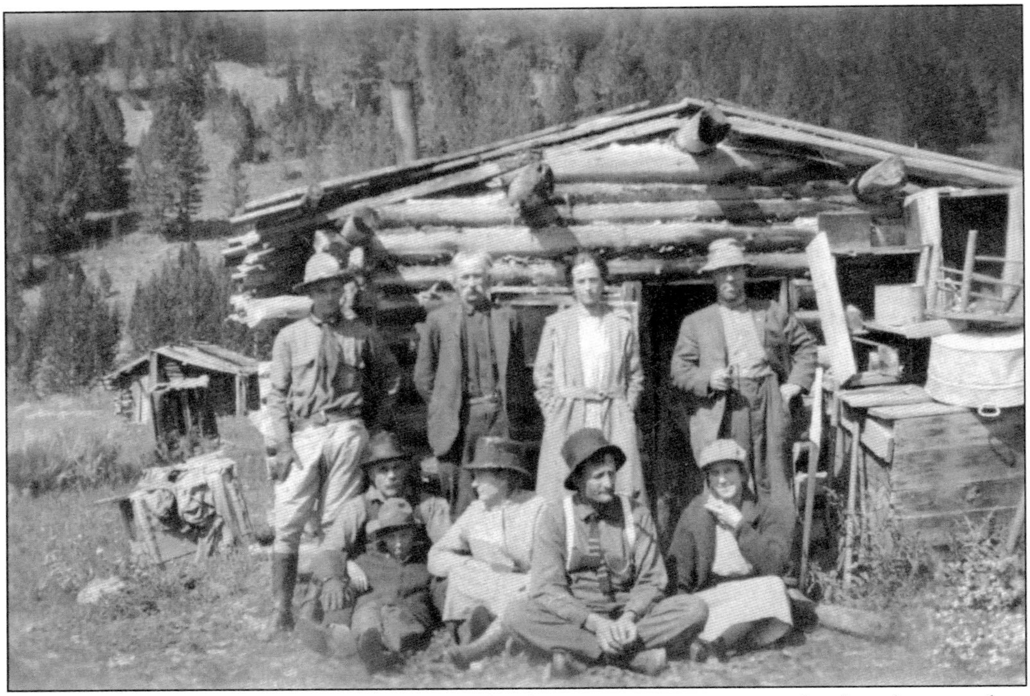

Life was harsh in this rugged country. Some hardy pioneers pause to pose for this photograph.

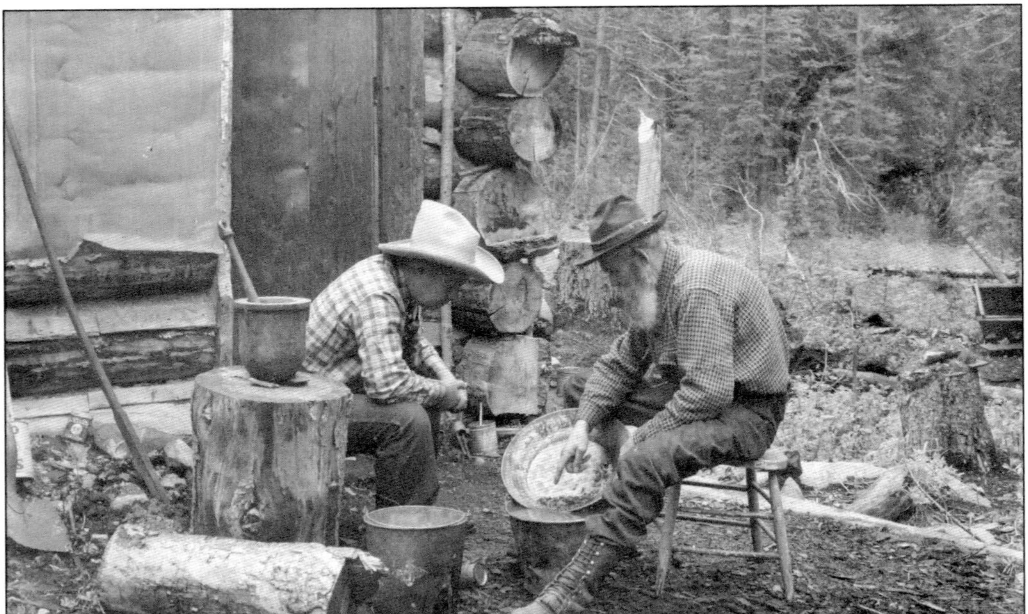

Placer mining brought attention to mineral finds and started some gold rushes. John "Jack" Thomas Murphy (right) is showing the gold he recovered by panning. Of note is his mortar for grinding rock on the chopping block.

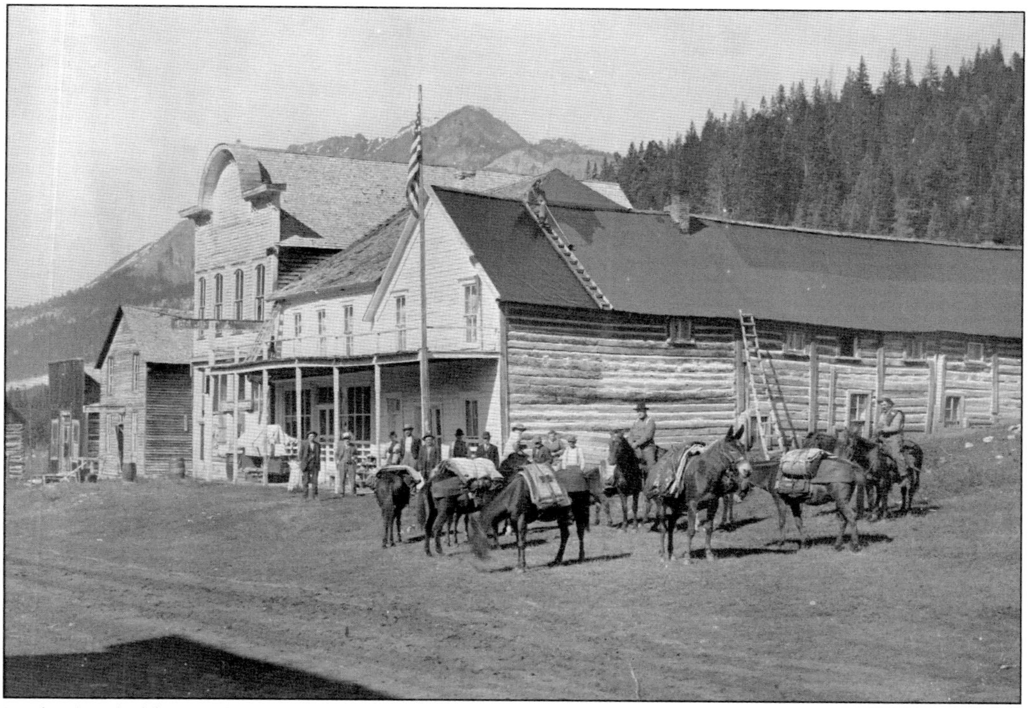

Mules hauled heavy loads where wagons could not go. Ore was hand-sorted from waste, packed into sacks, and taken by packhorses or mules into Cooke City. It was then shipped by wagon to Cinnabar and by train to East Helena.

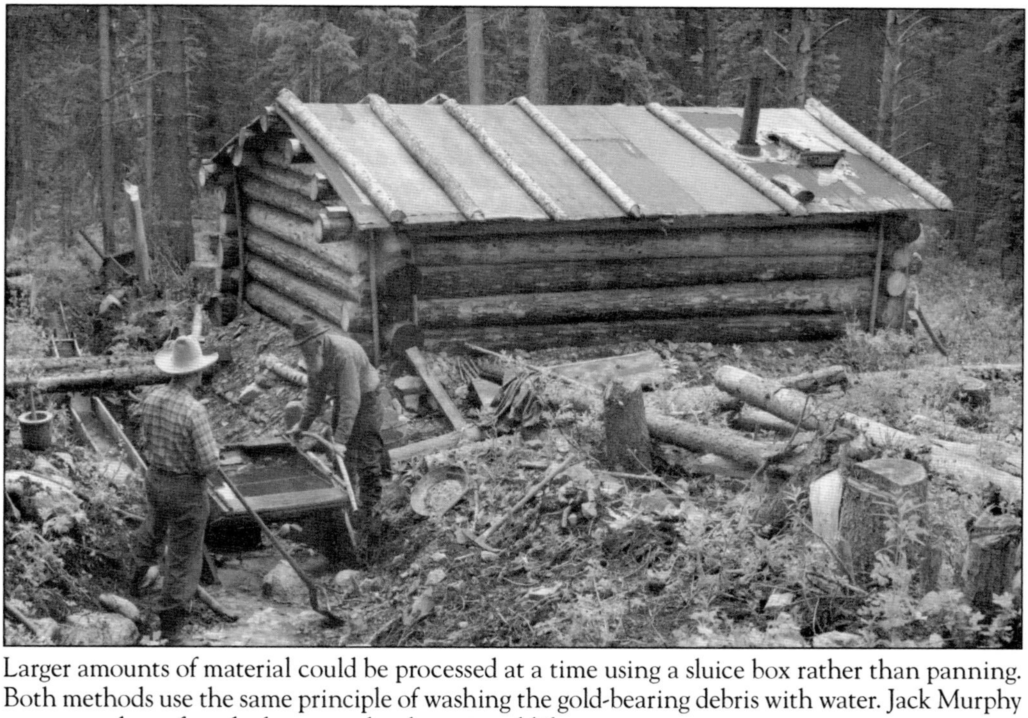

Larger amounts of material could be processed at a time using a sluice box rather than panning. Both methods use the same principle of washing the gold-bearing debris with water. Jack Murphy may never have found a bonanza, but he enjoyed life.

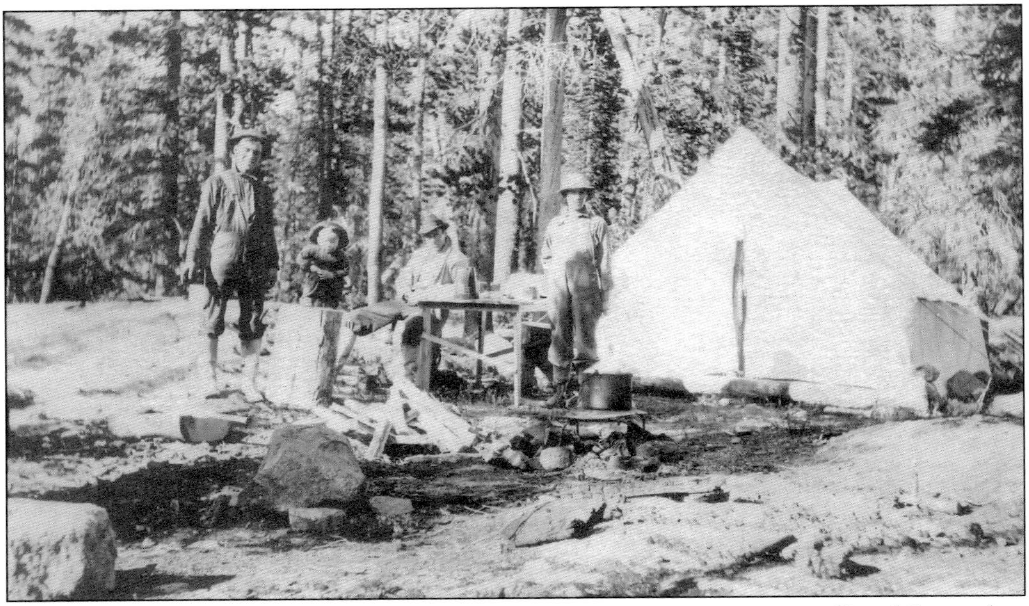

Dated September 1917, this photograph is of a tent that was at Broadwater Lake Camp.

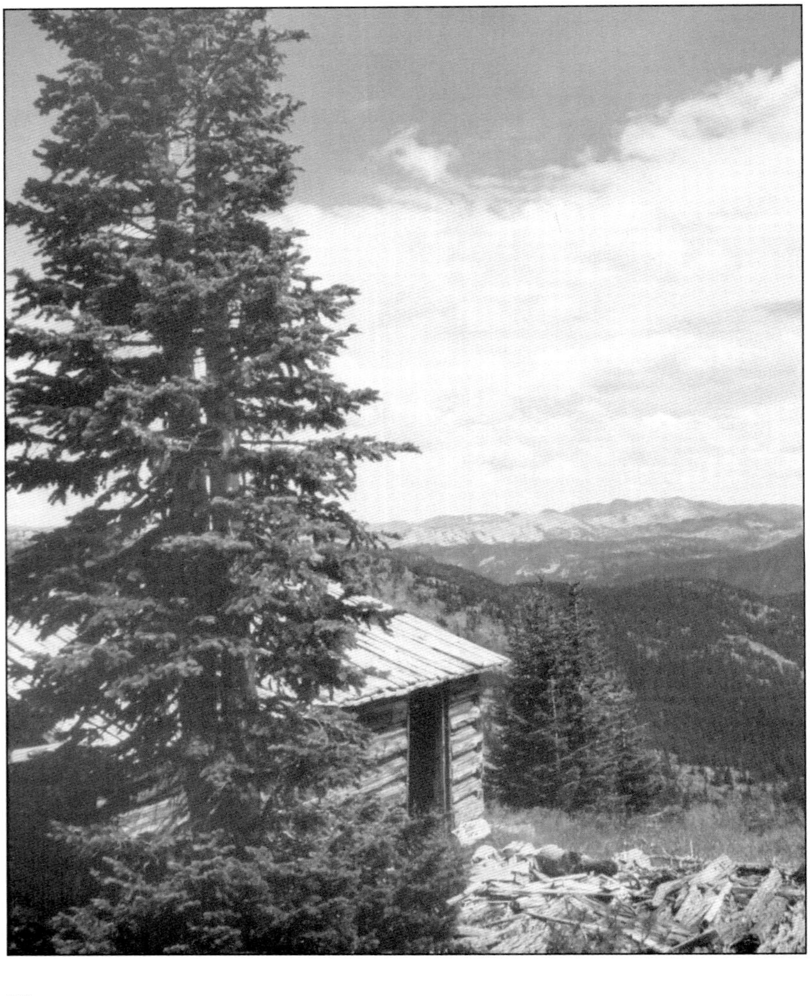

Miners were determined to protect their mines from claim jumpers, even with a gun. They built their cabins nearby, but some had a long way to go for tobacco.

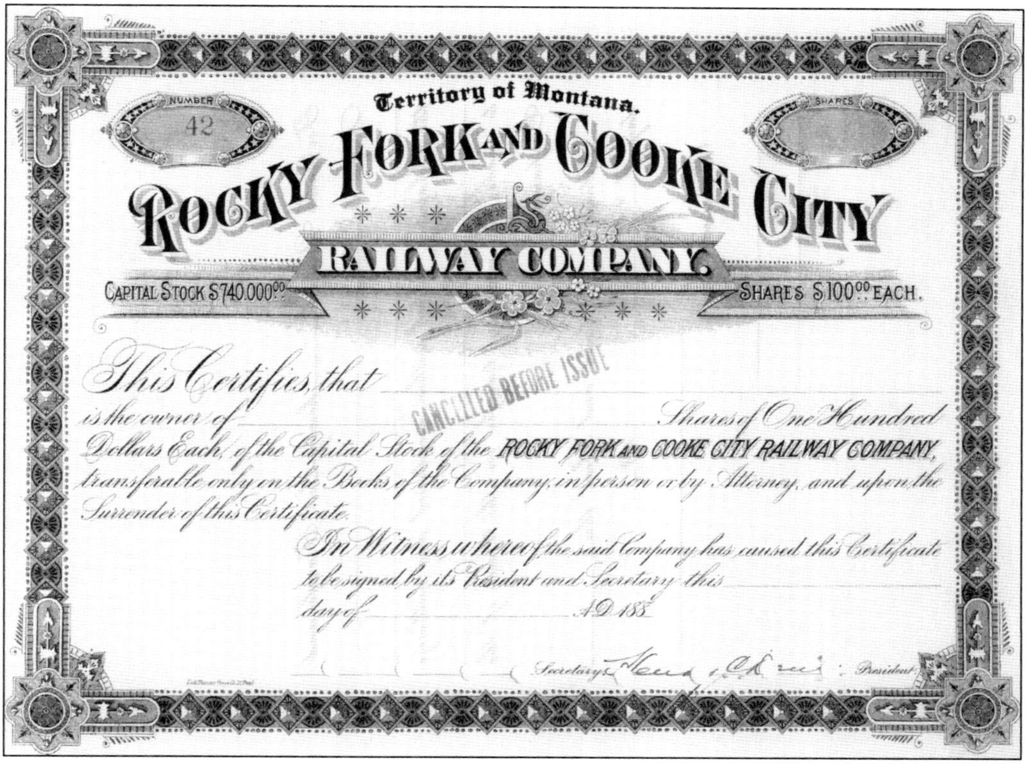

Territory of Montana.

NUMBER 42 SHARES

ROCKY FORK AND COOKE CITY RAILWAY COMPANY.

CAPITAL STOCK $740,000.00 SHARES $100.00 EACH.

This Certifies, that is the owner of Shares of One Hundred Dollars Each, of the Capital Stock of the ROCKY FORK AND COOKE CITY RAILWAY COMPANY, transferable only on the Books of the Company, in person or by Attorney, and upon the Surrender of this Certificate.

CANCELLED BEFORE ISSUE

In Witness whereof the said Company has caused this Certificate to be signed by its President and Secretary this day of A.D. 188

Secretary. President.

In 1886, entrepreneurs organized the Rocky Fork & Cooke City Railroad, proposing a line near Billings to the coalfields near Red Lodge and extending it toward Cooke City. It never progressed beyond dreams for Cooke City. This Rocky Fork & Cooke City Railroad stock certificate was stamped "cancelled" before it was issued. (Courtesy of David DePuy.)

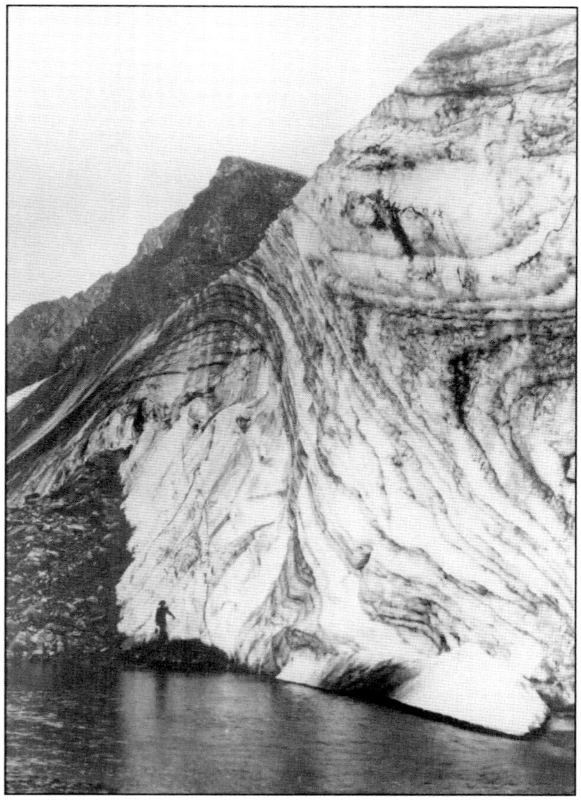

Martin Ranmael and Tom Christianson made one of the earliest visits to the famous Grasshopper Glacier about 1885 and found evidence of prospecting. Many thought the glaciers were only huge snowbanks, but a professor and a geologist from Bozeman confirmed that this was a glacier. Ranmael had a contract in the winter of 1886–1887, packing a ton of supplies, provisions, mining tools, powder caps, and fuses for a camp northeast of Cooke City. He carried from 75 to 120 pounds each trip and had many tales of his experiences.

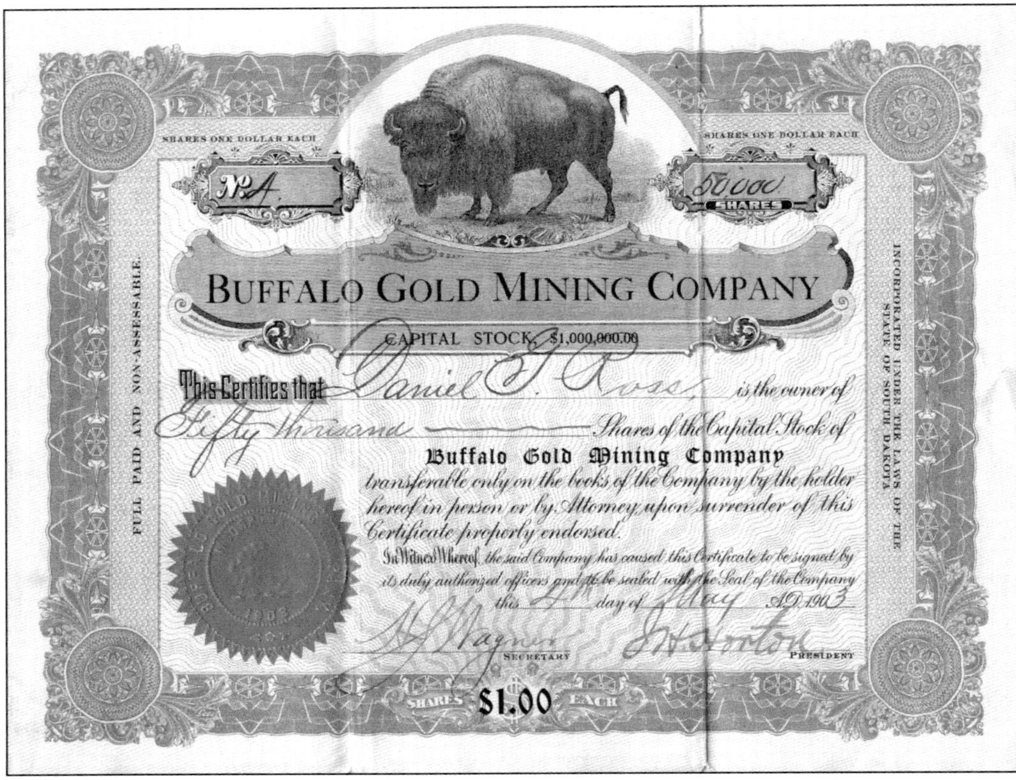

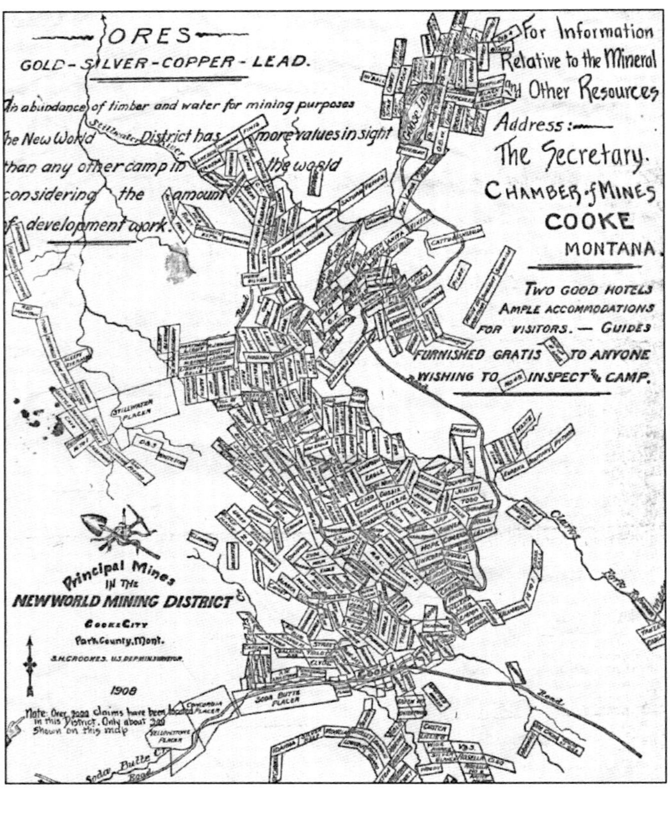

Buffalo Gold Mining Company was organized by R.I. McKay, and his associates bought the Republic Mine in 1906. The mine operated intermittently with little success. This stock certificate for 50,000 shares at $1 each is dated May 4, 1903. (Courtesy of David DePuy.)

An advertisement found on the back of a letterhead reads, "Principal Mines of the New World Mining District 1908. Note: Over 3,000 patented and unpatented claims had been located in the district. Only about 300 are shown on this map. Two good hotels—ample accommodations for visitors—guides furnished gratis to anyone wishing to inspect the camp."

Three

Cooke City Color

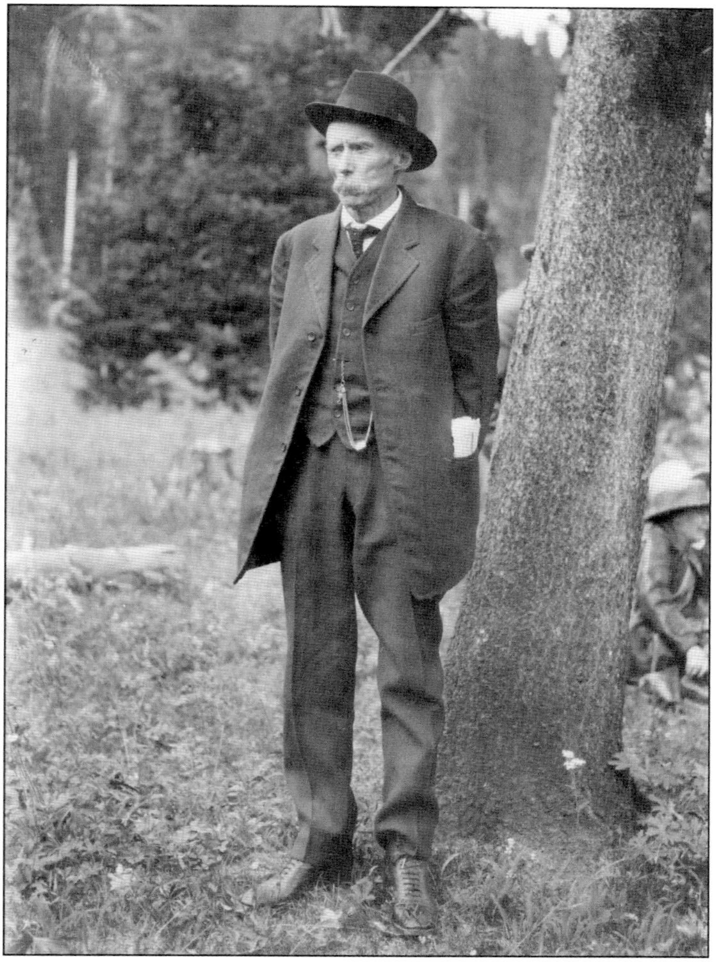

John "Jack" Allen is dressed in his black suit with distinct watch chain and is quoted as saying, "I traded in mining claims, too and put my profits in Cooke property. I remember in 1889 I needed a watch, so I traded a mine for one. Oh yes, I got some 'boot' in the deal." In April 1882, Jack Allen came to Cooke City with a four-horse team, building his own road through the mountains over 100 miles. There were no roads from Mammoth Hot Springs and only horseback trails up the Yellowstone, Lamar, and Soda Butte Creek. In 1884, Allen was the proprietor of the Cosmopolitan Hotel, a social center of the community. He lived in Cooke City more than 60 years and was still there when the scenic highway to Red Lodge was built.

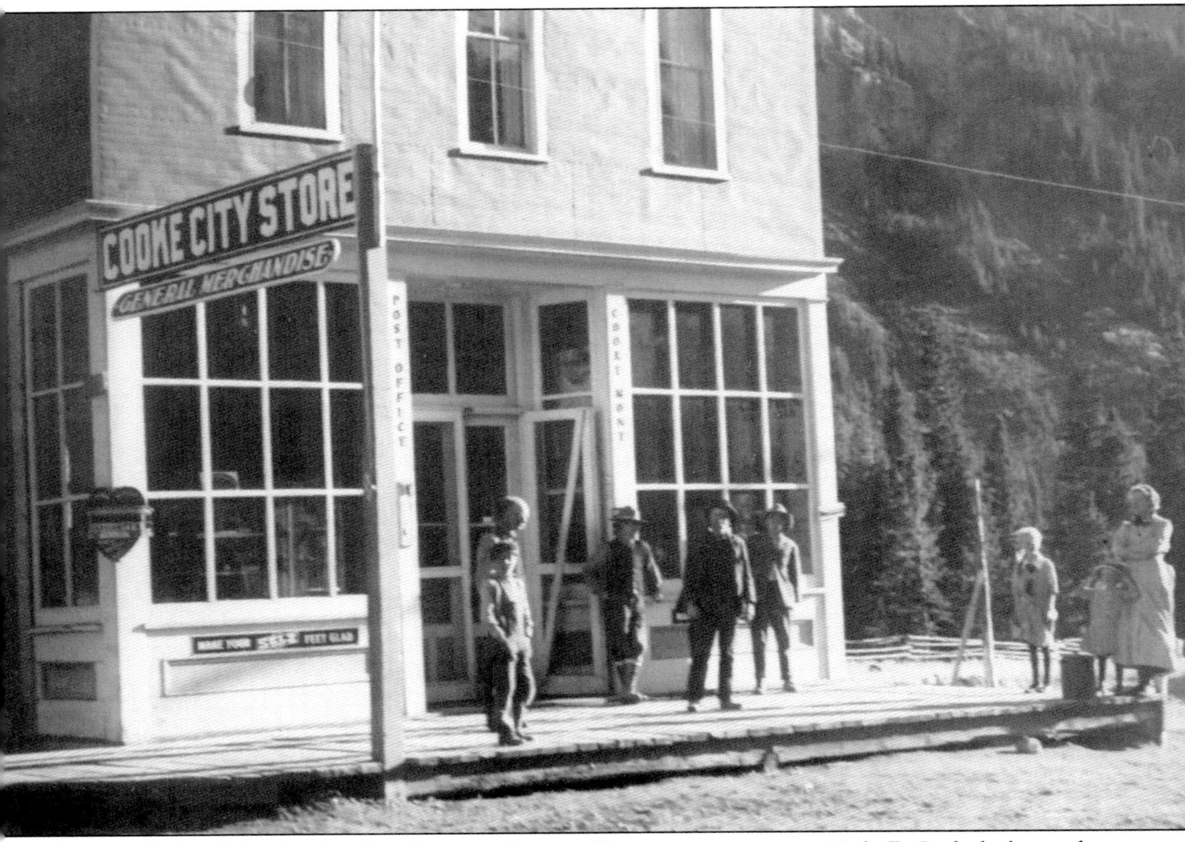

In front of the Cooke City Store and post office are store owners Nels E. Soderholm and son Harold on the left and Elizabeth J. Soderholm and daughter Elizabeth Pauline on the right. The money earned by the Soderholms from the postal work enabled them to remodel part of the upstairs for an apartment. During the winter when Nels was postmaster, he would ski over Daisy Pass to meet a postal employee from Nye who stayed in a cabin near Lake Abundance where he delivered and retrieved the mail. Nye was 40 miles north. The store added new lines to its expanding inventory and continued to prosper. Early in 1911, the competing Allison Mercantile Company went bankrupt. Many of the ornate fixtures that had originally been put into the Cooke City Store when Allison had occupied it were then sold to Soderholms at a sheriff's sale and placed back in the store.

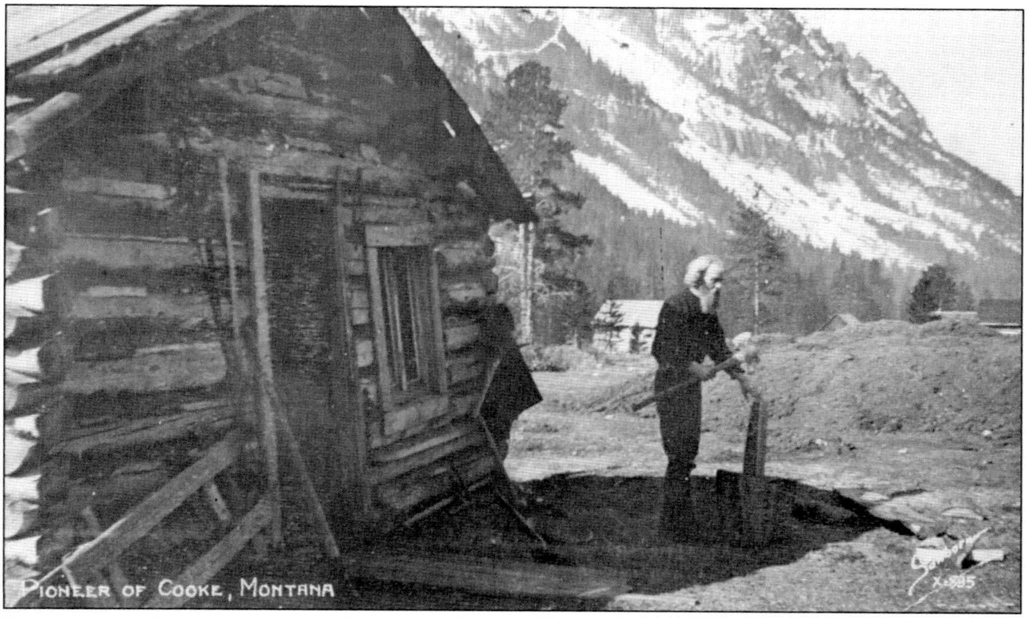

Deep snows of winter drove residents out of town. Some miners walked or skied to Red Lodge to work in the coalfields for a grubstake. Those who stayed needed to have plenty of provisions and wood stacked. Warming himself twice, this old-time pioneer of Cooke City is chopping wood.

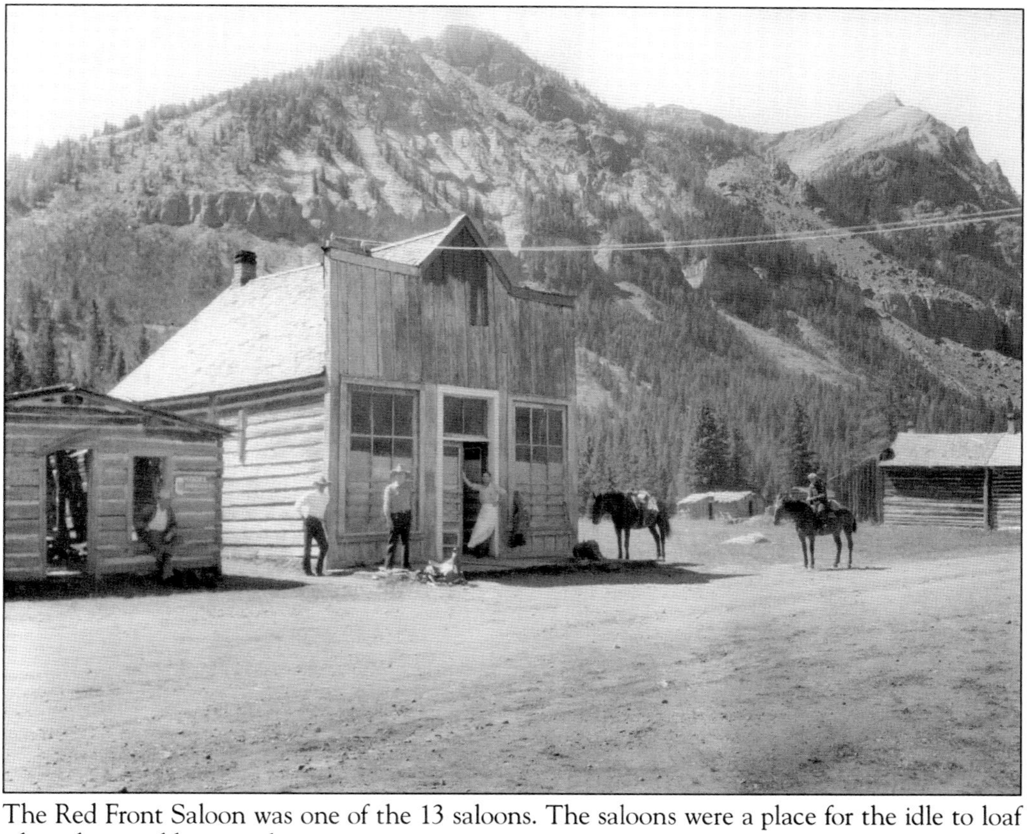

The Red Front Saloon was one of the 13 saloons. The saloons were a place for the idle to loaf when they could not work.

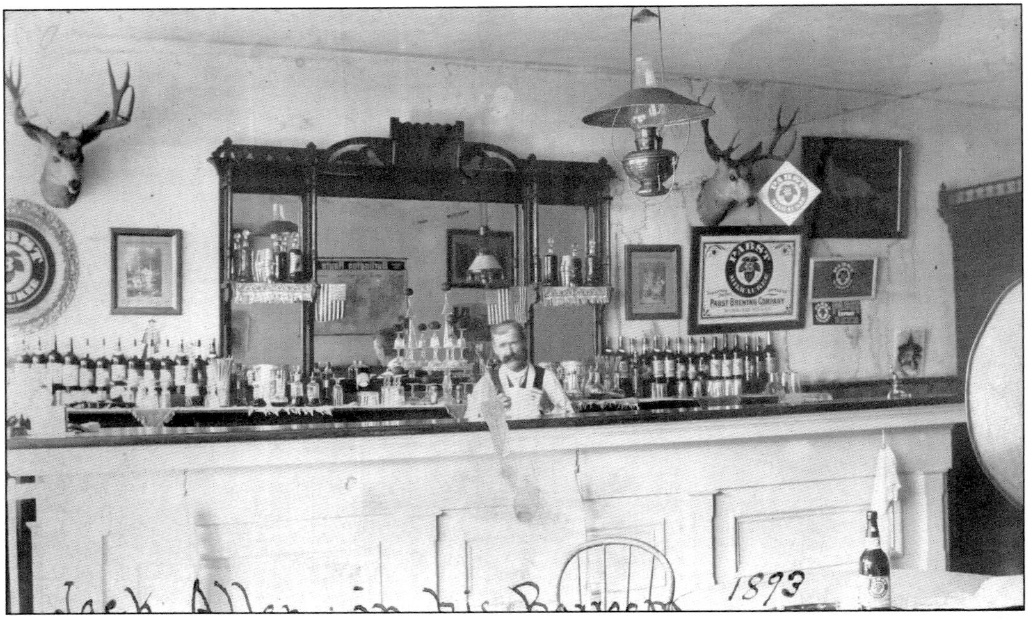

Jack Allen is pictured in his barroom in 1893. Pabst Milwaukee was the brew of choice. The reflection in the mirror is a Burlington Route map advertisement. In 1885, a high-stakes cribbage match was played at the Cosmopolitan waging 100 feet of one claim for 100 feet of another.

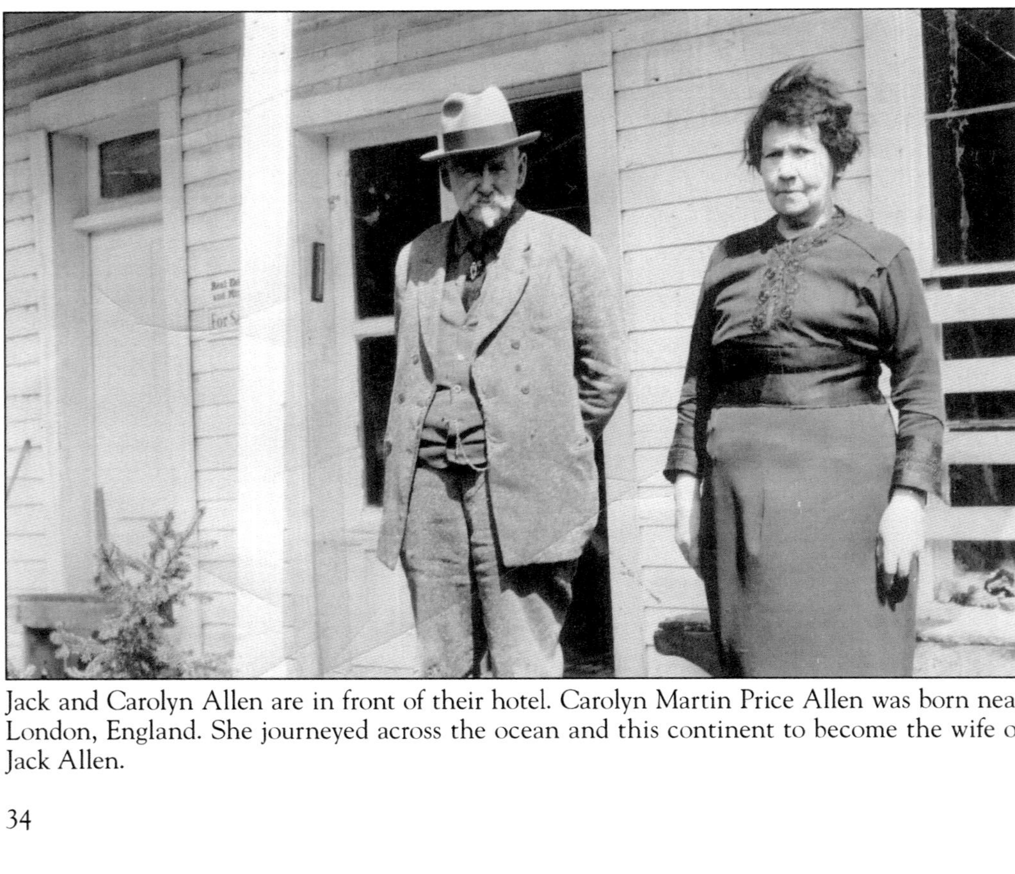

Jack and Carolyn Allen are in front of their hotel. Carolyn Martin Price Allen was born near London, England. She journeyed across the ocean and this continent to become the wife of Jack Allen.

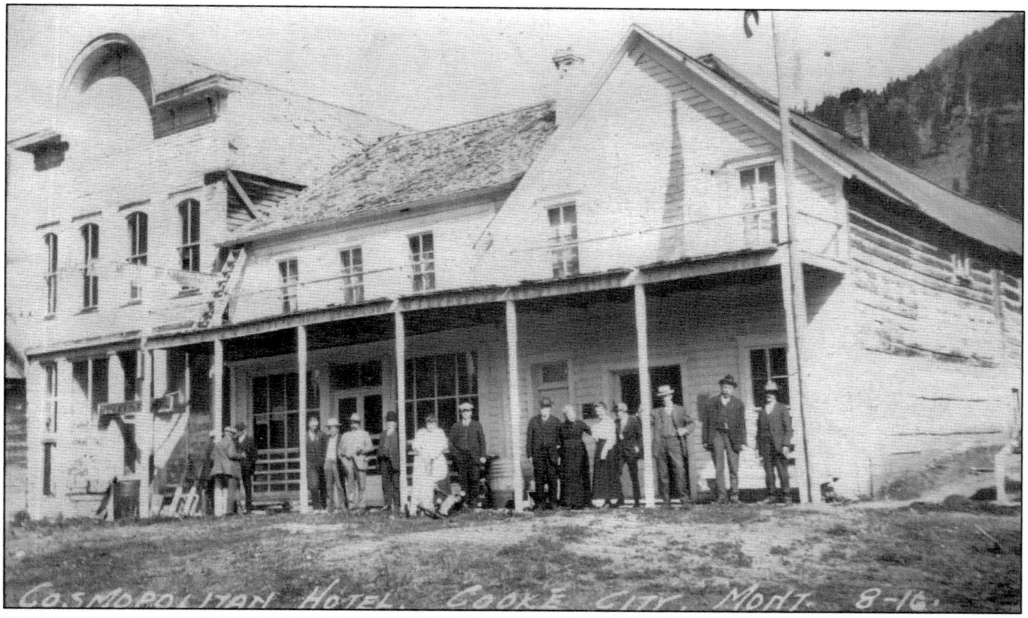

Pictured in front of the double doors at the Cosmpolitan Hotel, fifth from left, is Dr. G.L. Tanzer. The date appears to be August 1916. Tanzer emigrated from Germany to the United States in 1885. He was appointed chemist for Seattle and the state of Washington. He had valuable holdings near Yellowstone National Park.

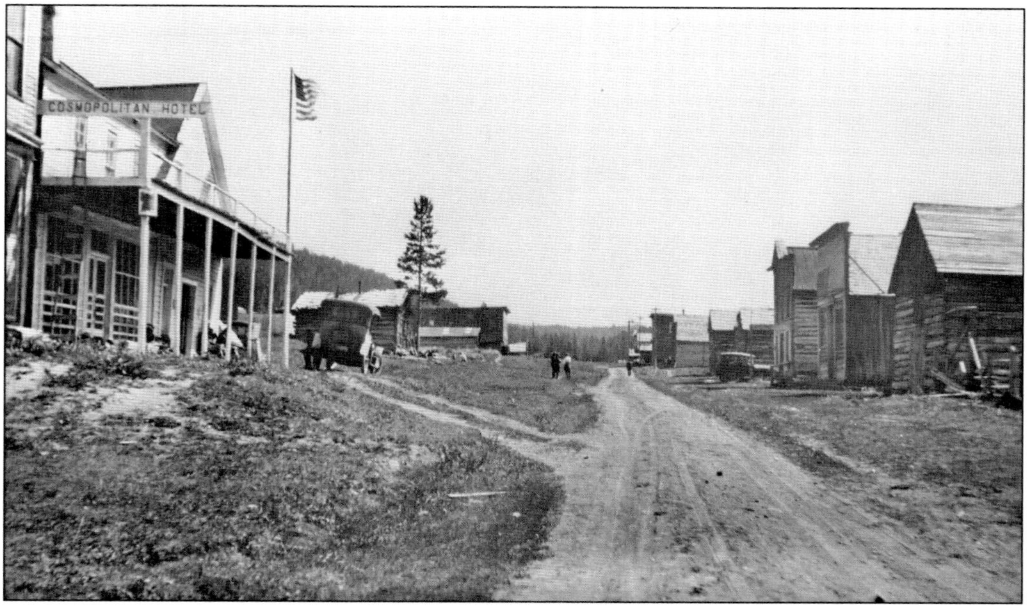

Looking east, the Cosmopolitan is on the left, and on the right is the A.O. Saloon, next door to the Curl Hotel. The Cosmopolitan and Curl Hotels were boardinghouses for many bachelor miners. When the grade of Main Street was changed for the new scenic highway, the Cosmopolitan sat high above the street.

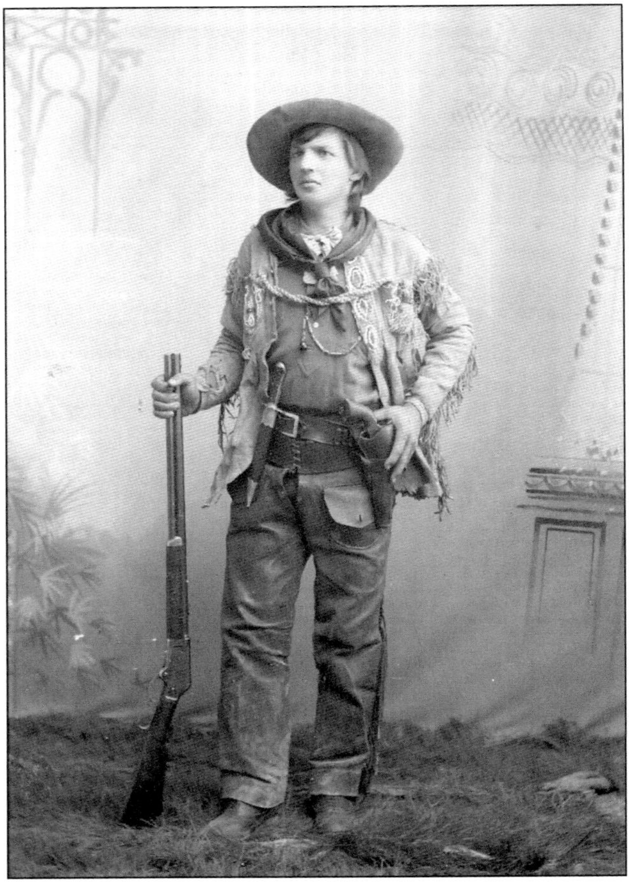

This vintage photograph is of George Reeb, also known as "Morphine Charley." Reeb and "Little Gus" Smitzer robbed a stagecoach near the old Canyon Hotel in Yellowstone National Park on August 14, 1897, and Reeb was later convicted for highway robbery. After serving two years and a month in prison, he returned to Judge John Meldrum, thanking him for the help to break his morphine habit. He became a productive citizen, working in Butte and Cooke City mining. He was supervisor for the Buffalo Montana and Glengarry Mining Companies and developed the Henderson/Last Chance claims with his own capital.

Calamity Jane frequented Cooke City and was one woman who was allowed to gamble in the saloons. Calamity Jane and her brother were acquaintances of George Reeb. Other prominent people were Ernest Hemingway, Charlie Russell, Fredrick Lowe, and Mary Robert Rinehart.

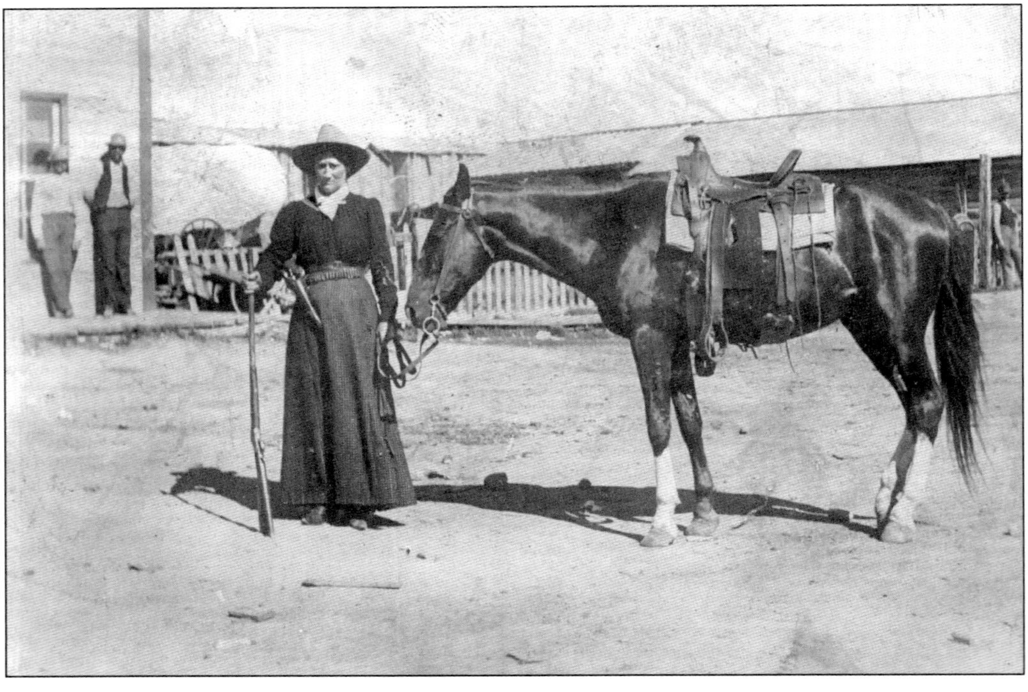

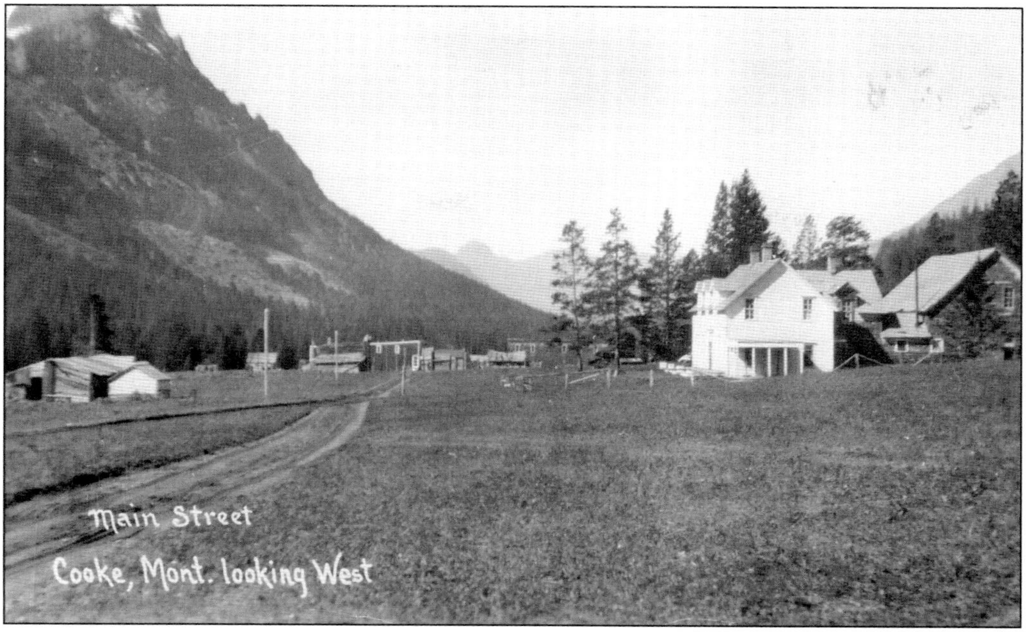

This postcard reads, "Main Street Cooke, Mont. looking west." George O. Eaton, of the Republic Mine, was an easterner and built this house on Main Street. The architecture of the whiteboard siding is typically New England and was unlike the crude shacks and buildings surrounding it. The house was later owned by Robert McKay and still is a landmark on Main Street.

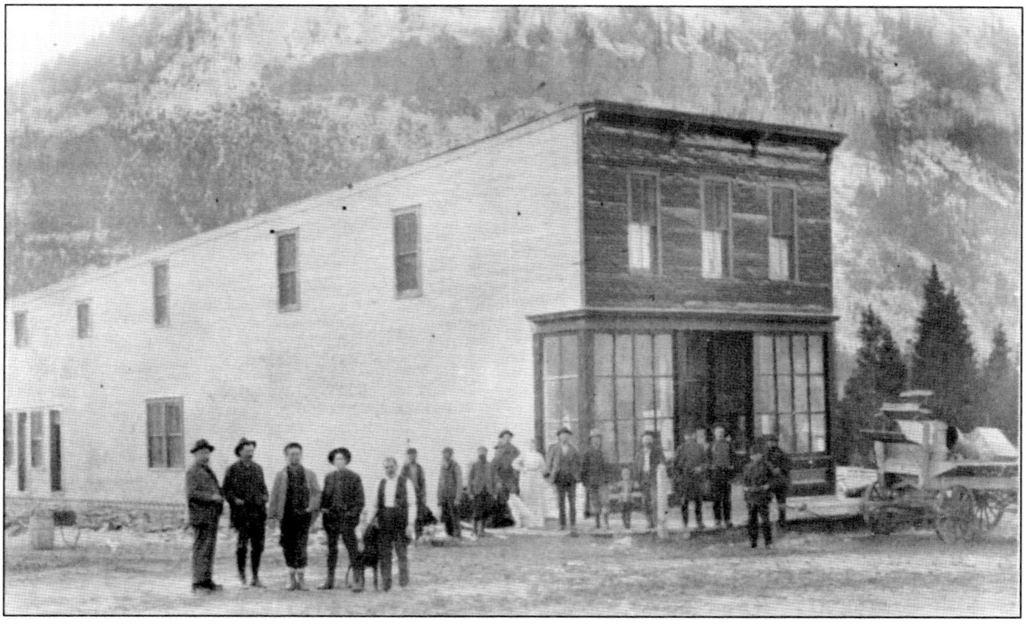

The Cooke City Store celebrated its 100th anniversary on March 27, 1986, and is listed in the National Register of Historic Places. Ralph Glidden said, "It has retained excellent historic architectural integrity and is highly unusual as an intact, historically functioning small town mercantile store."

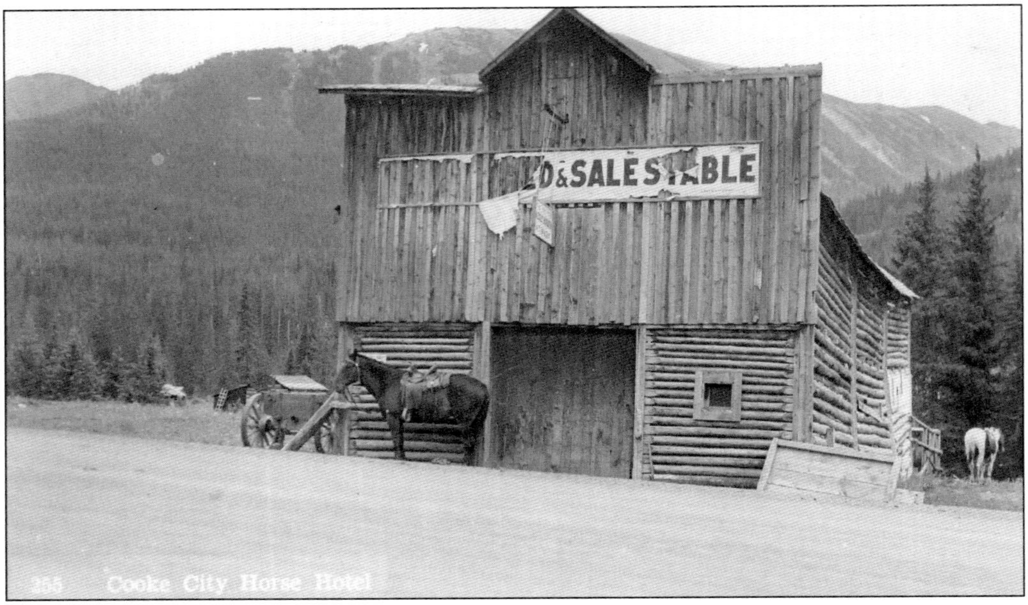

Shown is the local horse hotel and sales building. Travel was on horseback before the roads were improved. They were used to haul supplies and ore and to pull wagons.

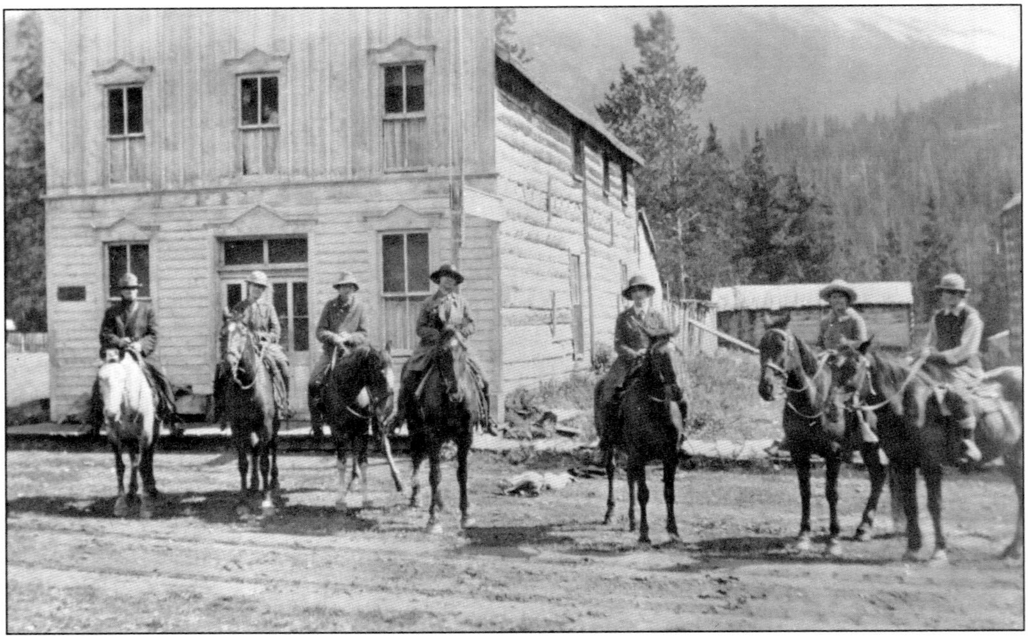
Mounted riders pose in front of the Curl House. No sidesaddles were needed for these hardy women.

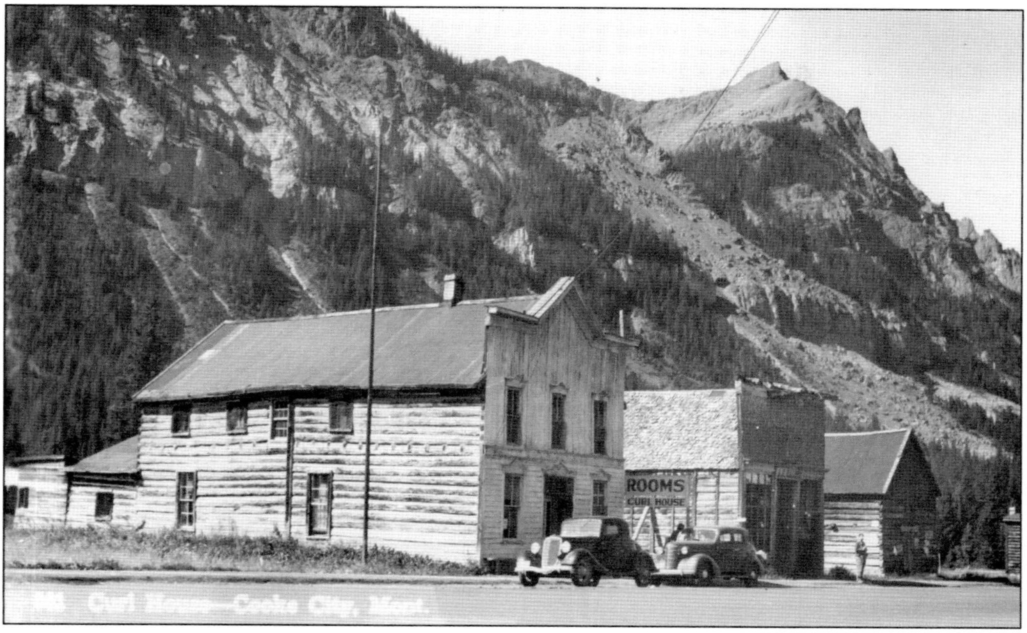

The Curl House was a hotel and boardinghouse for miners owned by John F. and Zona Curl. The A.O. Saloon was next door. The better bars vied for customers.

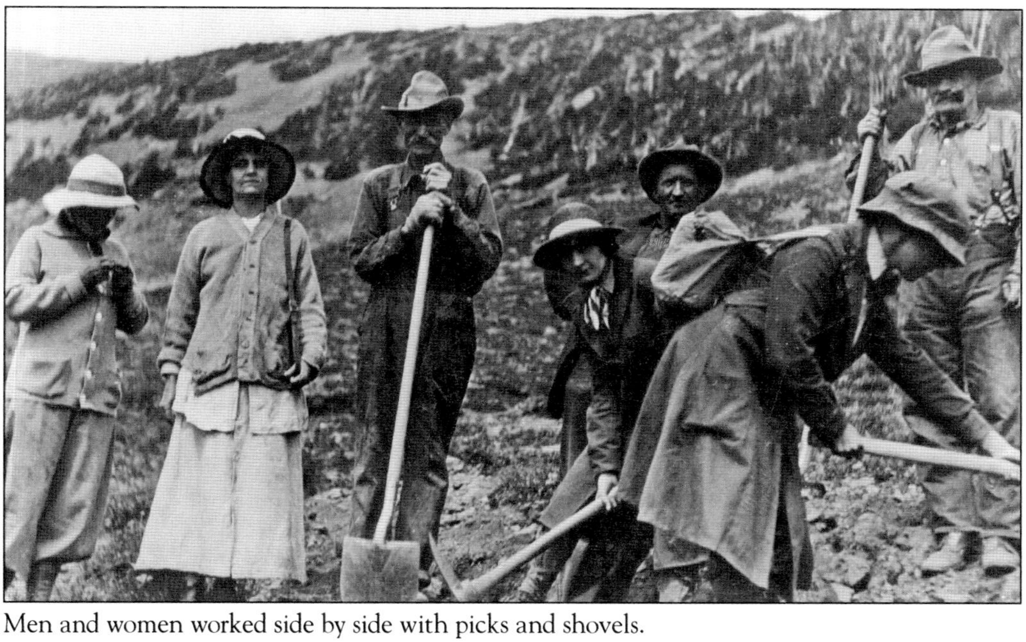
Men and women worked side by side with picks and shovels.

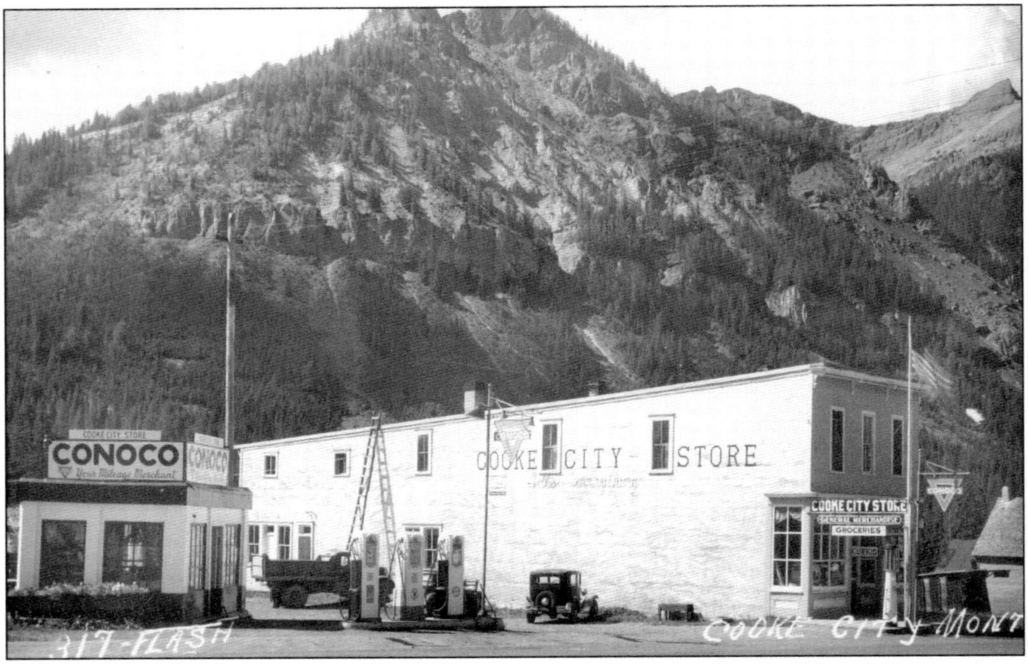

In anticipation of the automobile tourist, Nels Soderholm purchased a bulk fuel tank and hauled gasoline to a Conoco station he operated adjacent to the store. A sign advertises it as "Conoco Station Your Mileage merchant." The *Carbon County News Red Lodge* reported that gas stations sprang up, and the Red Lodge–Cooke City "high gear" road was more than an engineering achievement; it was the top-of-the-world road. It provided a fifth entrance to Yellowstone National Park.

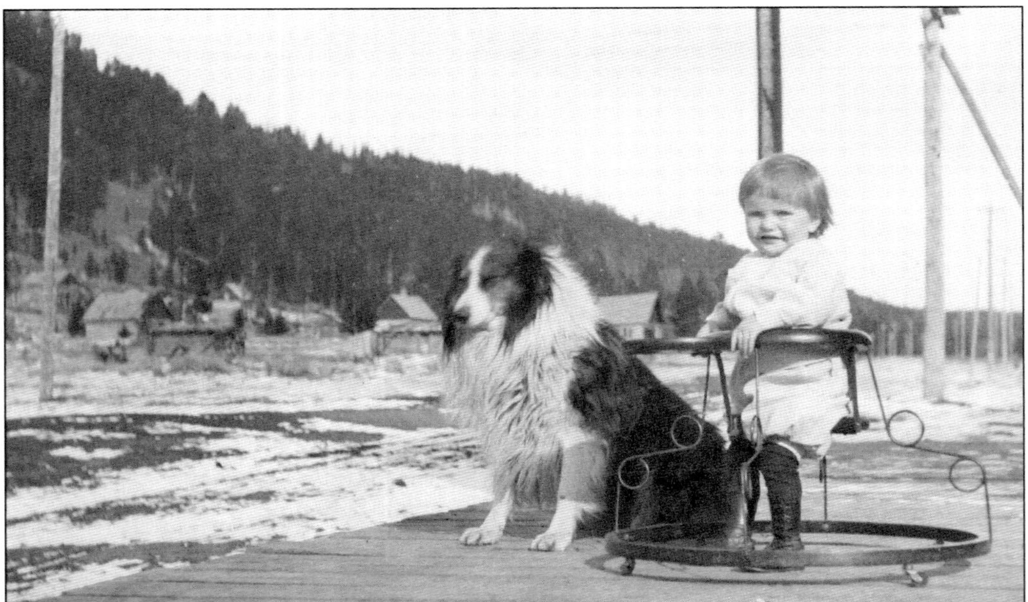
The greeting on this card is, "Merry Christmas Dec, 25, 1912 Betsy Soderholm." Elizabeth (Pauline) is with a canine friend in front of her parents' store.

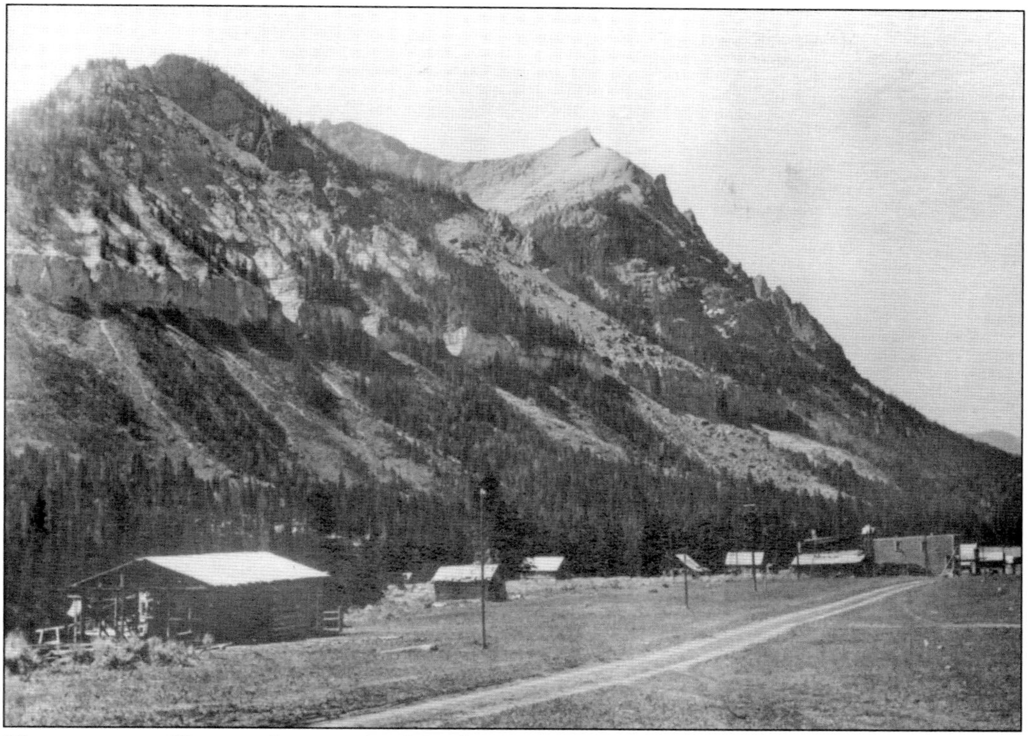

This image is a Haynes photograph of Cooke City.

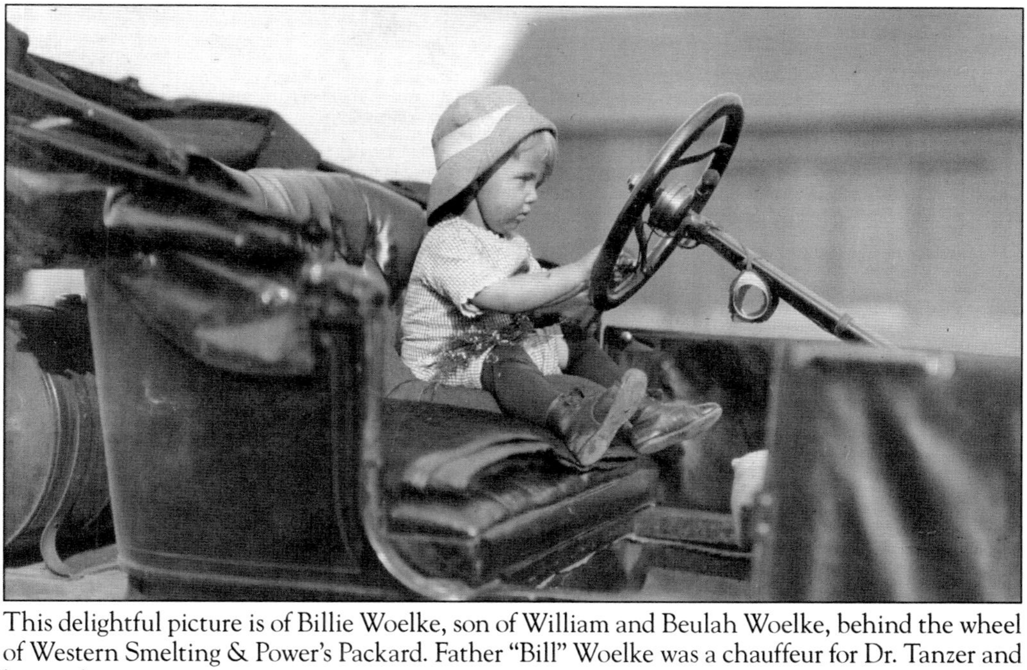

This delightful picture is of Billie Woelke, son of William and Beulah Woelke, behind the wheel of Western Smelting & Power's Packard. Father "Bill" Woelke was a chauffeur for Dr. Tanzer and his nephew.

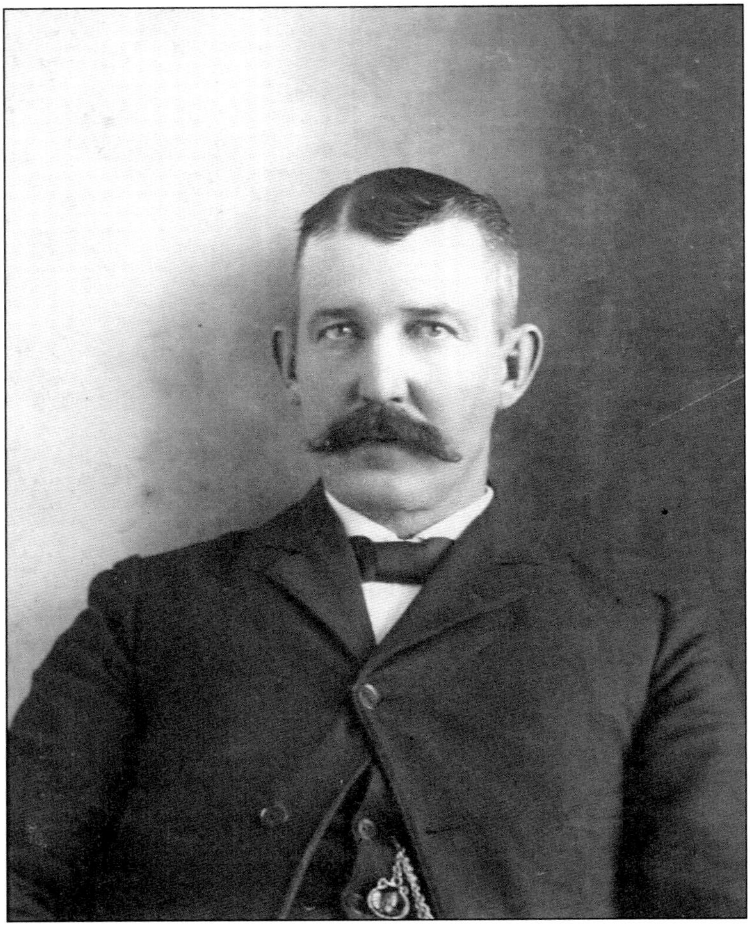

Thomas Harden died February 25, 1922. Harden had paid $200 for the homestead of Enos Kersey, for whom Kersey Lake was named. Kersey raised hay and sold it in town.

Written on the photograph is, "The only time you see mother earth is when you dig a grave. On road to Cooke not for me." From left to right are Kent, Helmar, and Mason.

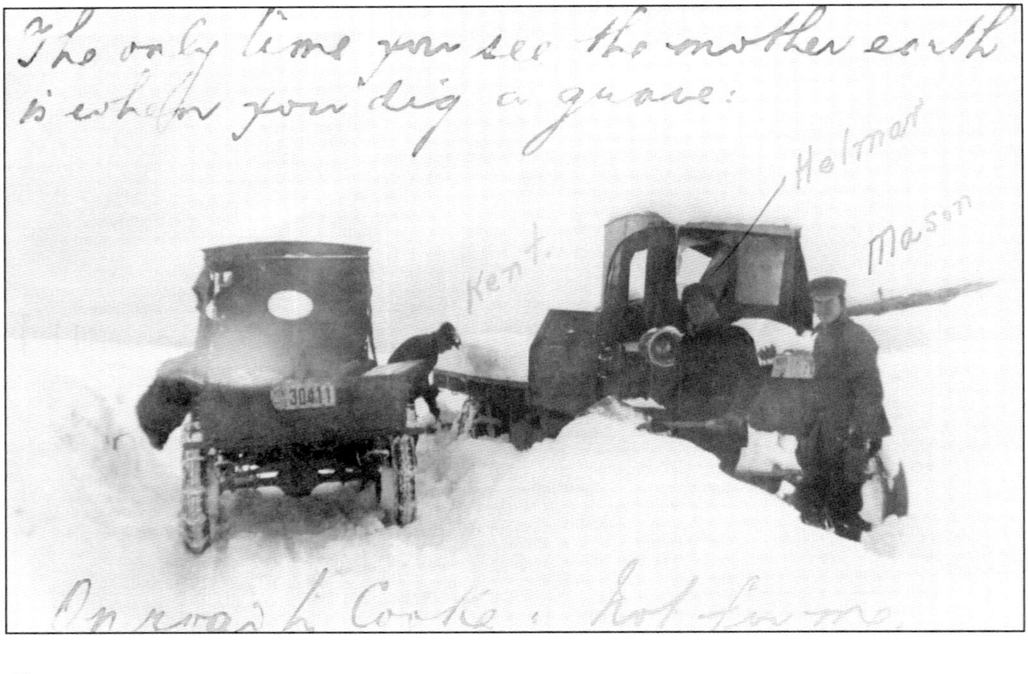

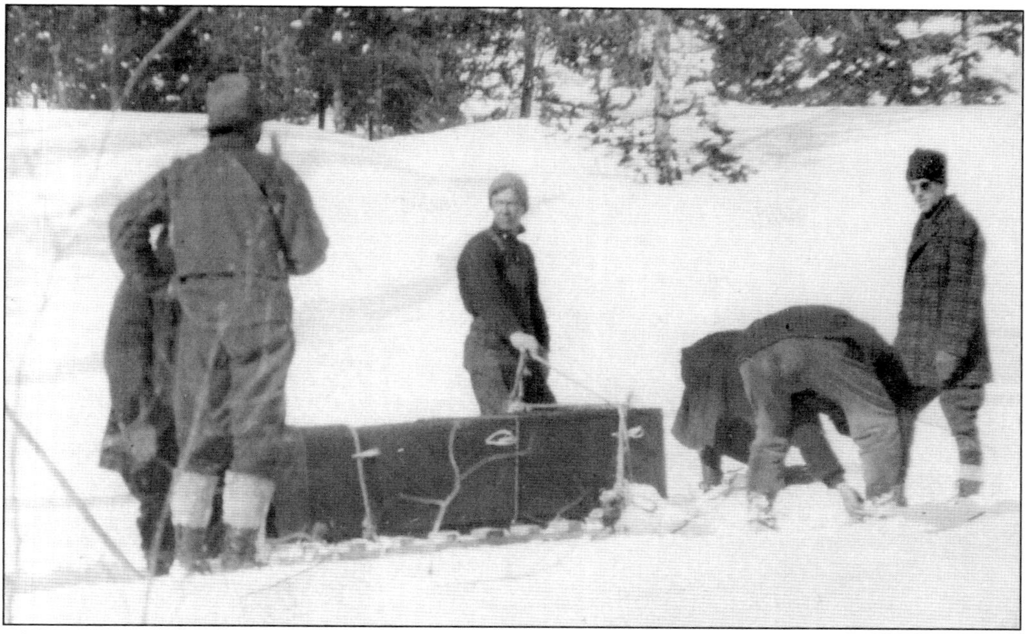

This photograph was taken on February 28, 1922, near the Allison smelter as men of the community were taking Thomas Harden's coffin for burial. When necessary, boards from uninhabited buildings were used to construct coffins. Harden's grave is marked by the only ornate grave fence in the Cooke City Cemetery.

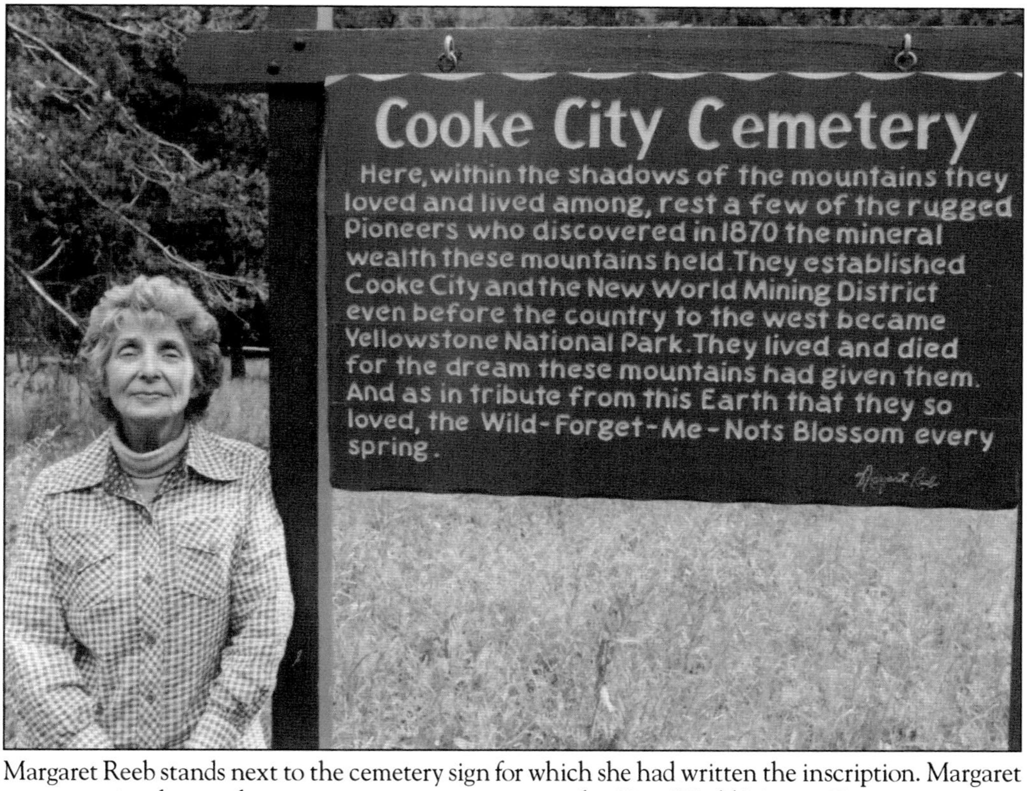

Margaret Reeb stands next to the cemetery sign for which she had written the inscription. Margaret was committed to exploring mining opportunities in the New World Mining District.

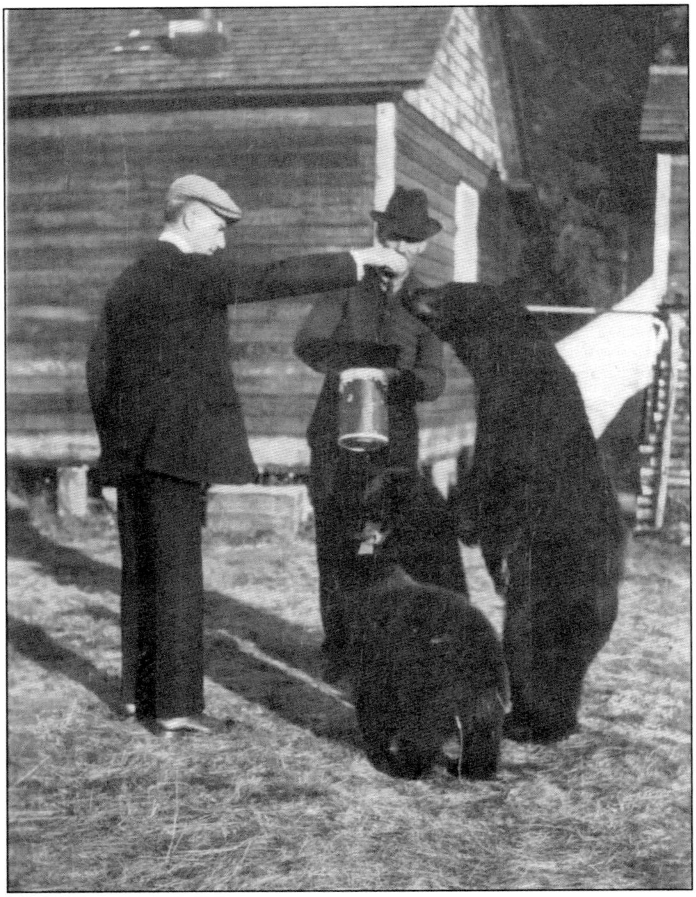

These men are dangerously feeding a mother bear and her cubs. This certainly is not something to do today.

Bill Tanzer sent this postcard of Index and Pilot to his sister Freda. It is postmarked September 26, 1914. The message reads, "They figure on stopping the stage the first of October. Won't that be a picnic? We will certainly be cooped up here. I climbed this mountain a couple weeks ago and she's a corker. Last week we had 18 inches of snow."

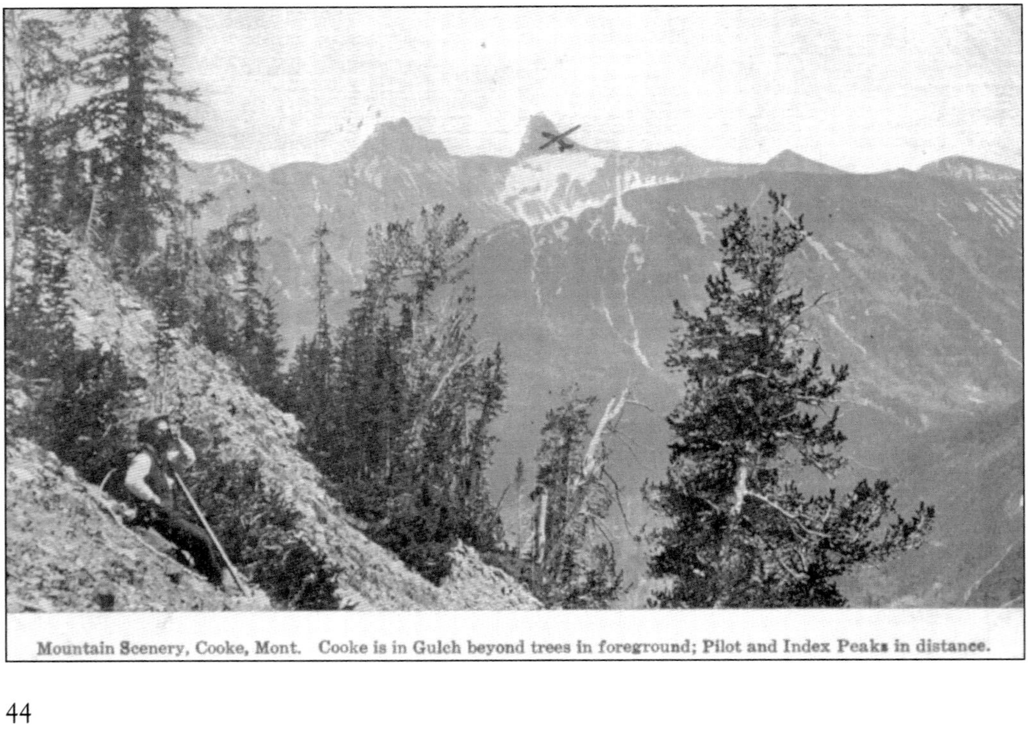

Mountain Scenery, Cooke, Mont. Cooke is in Gulch beyond trees in foreground; Pilot and Index Peaks in distance.

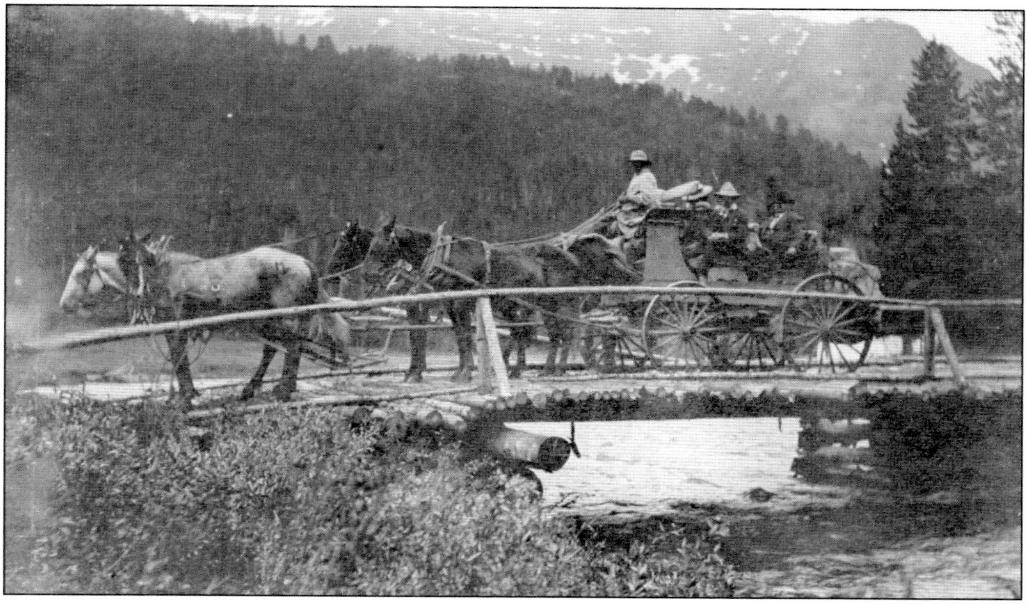

Traveling in the early days was made less difficult with bridges. C.J. "Jack" Barronette built the first bridge in the area in anticipation of mining. Bridges assured a dry crossing. According to an 1885 *Billings Herald*, 10 men volunteered to build a road into Cooke City from the south in exchange for grub and tools amounting to $400.

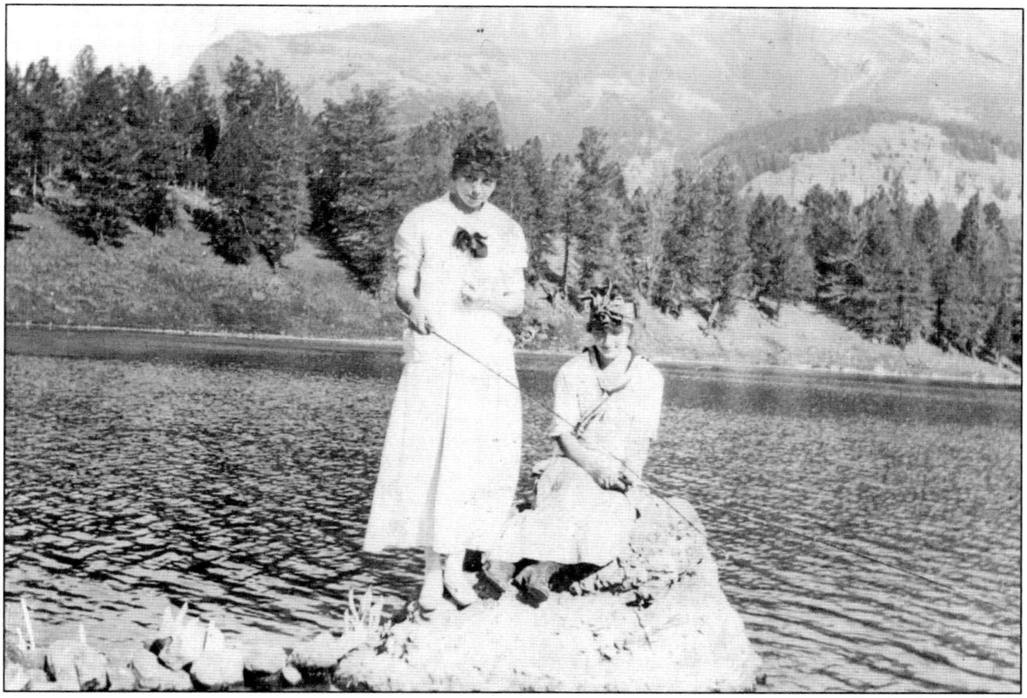

On a hot afternoon, the Tanzer family and friends loaded into the Packard for a drive and went to Fish Lake. Dr. Tanzer's daughters Alice (left) and Ruth are fishing in the lake.

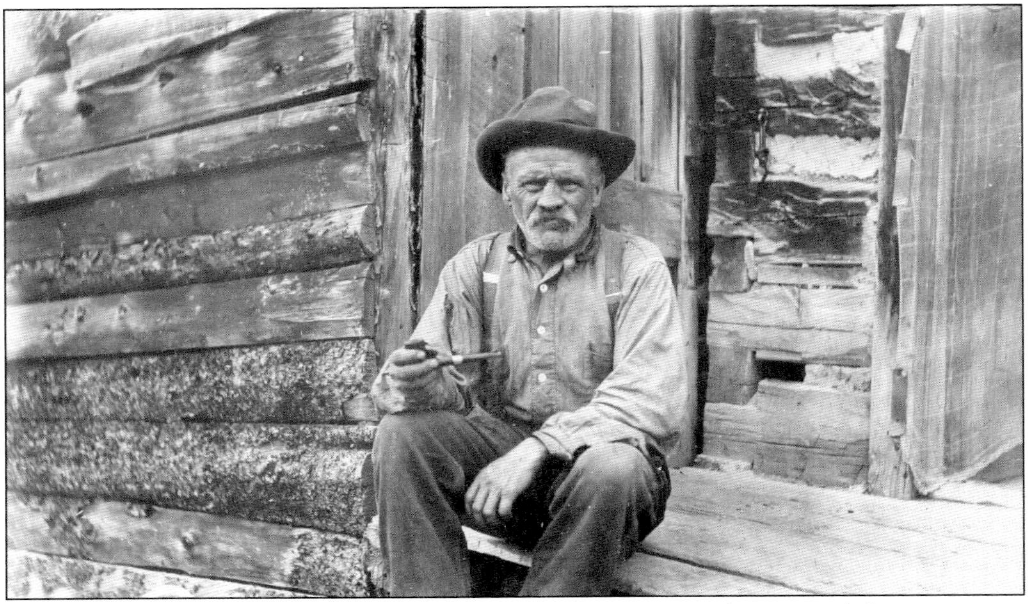

A sharp-witted character, Gene Underwood was the scion of the well-known Underwood family of Boston.

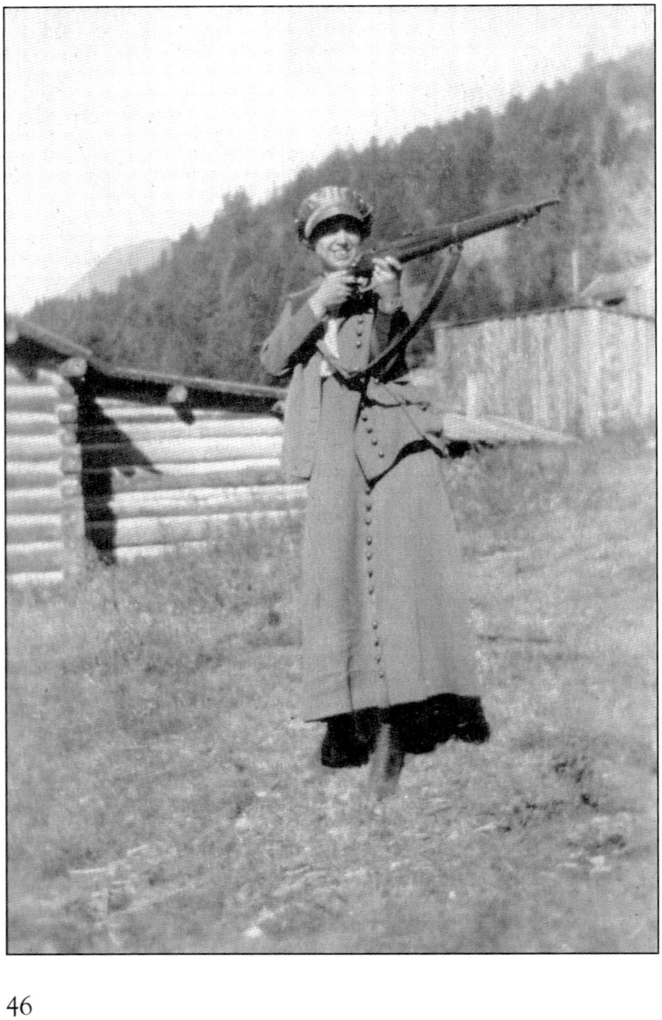

This is not Calamity Jane, but this young woman is learning how to shoot a gun.

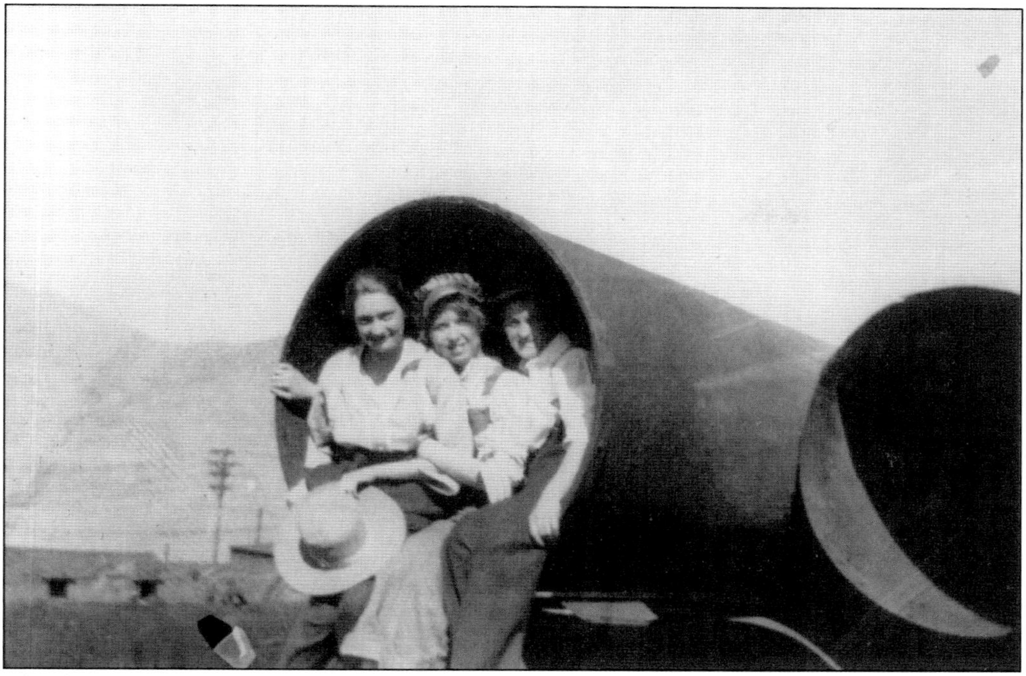

The three young women sitting in a part of the smelter stack give an idea as to the size of it.

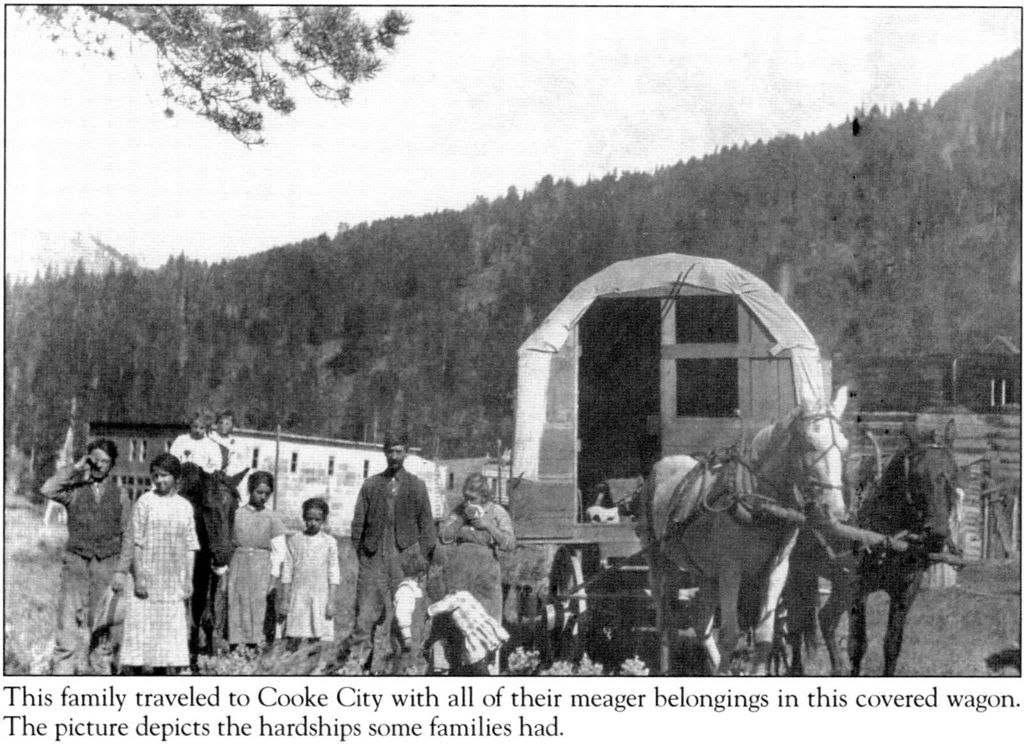

This family traveled to Cooke City with all of their meager belongings in this covered wagon. The picture depicts the hardships some families had.

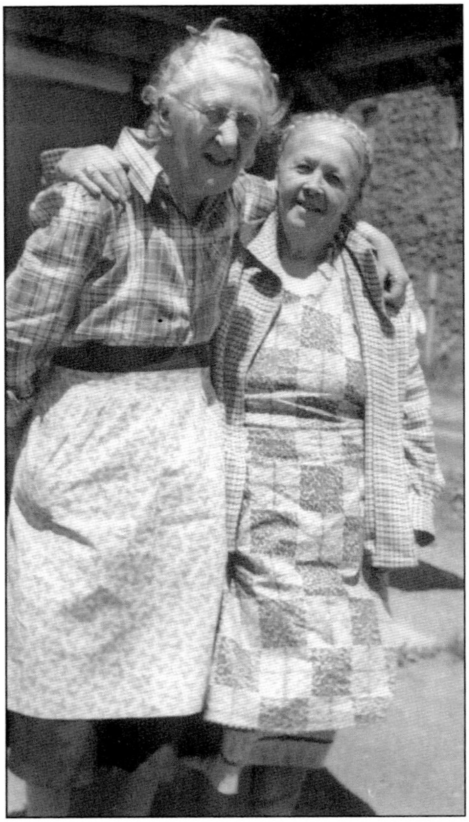

Early pioneers Elizabeth Soderholm (left) and Ingeborg Reeb embrace in front of the Cooke City Store in 1951. The tabulating pencil in Soderholm's hair was typical of her many storekeeper's duties for over a half a century and could indicate she was working at the store. Her nephew Sam Brady helped after Nels's death and took over the store after his "Auntie Lizzie's" retirement.

Thomas Blackhart was secretary of the Alice E. Mine in 1889 and built this hand-hewn log house. Three of Ingeborg Holland Reeb's children were born here.

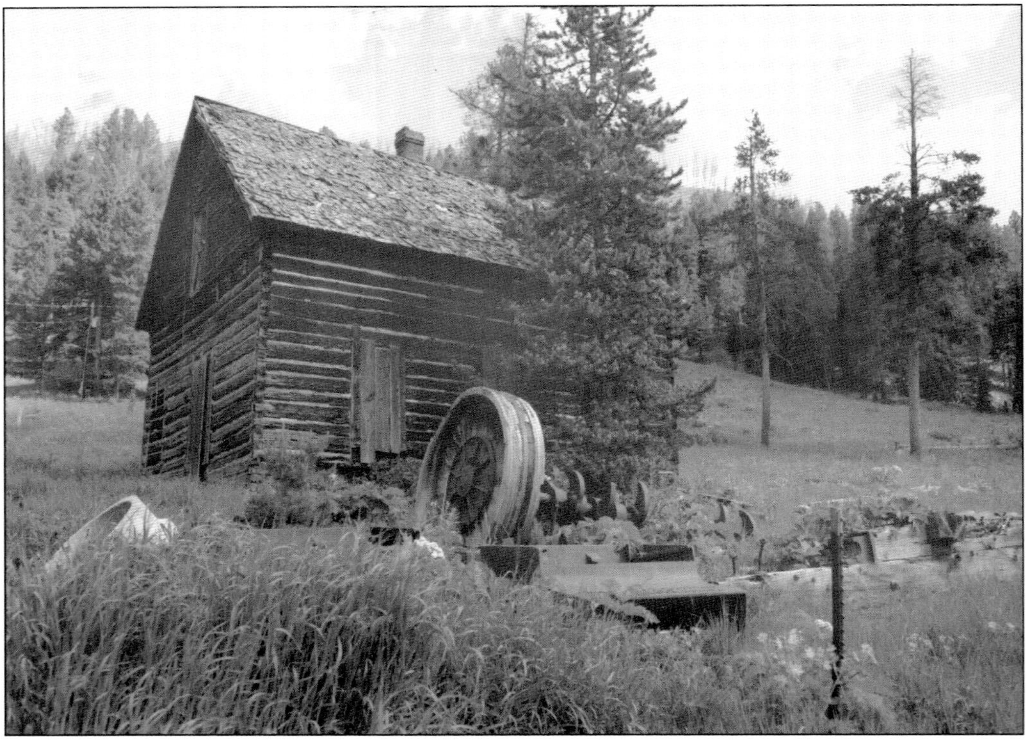

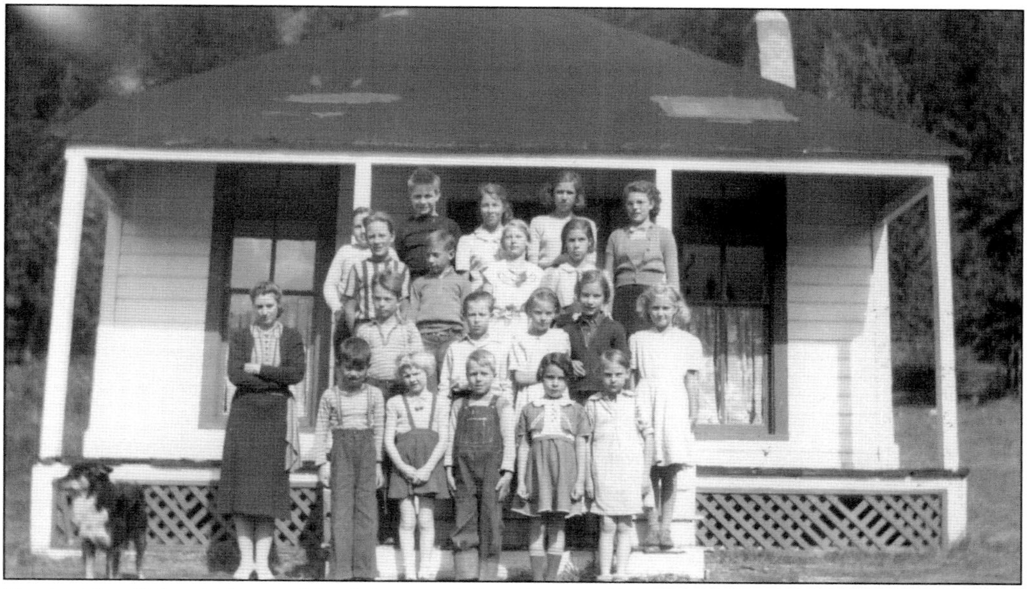

Cooke City School District No. 9 is one of the oldest in Park County. In this 1940–1941 school year photograph, teacher Grace Nye is pictured with 19 students. Pictures of an earlier school building showed two sets of stairs, which was a Swedish tradition. As the snow got deeper, fewer steps had to be climbed. (Courtesy of Yellowstone Gateway Museum.)

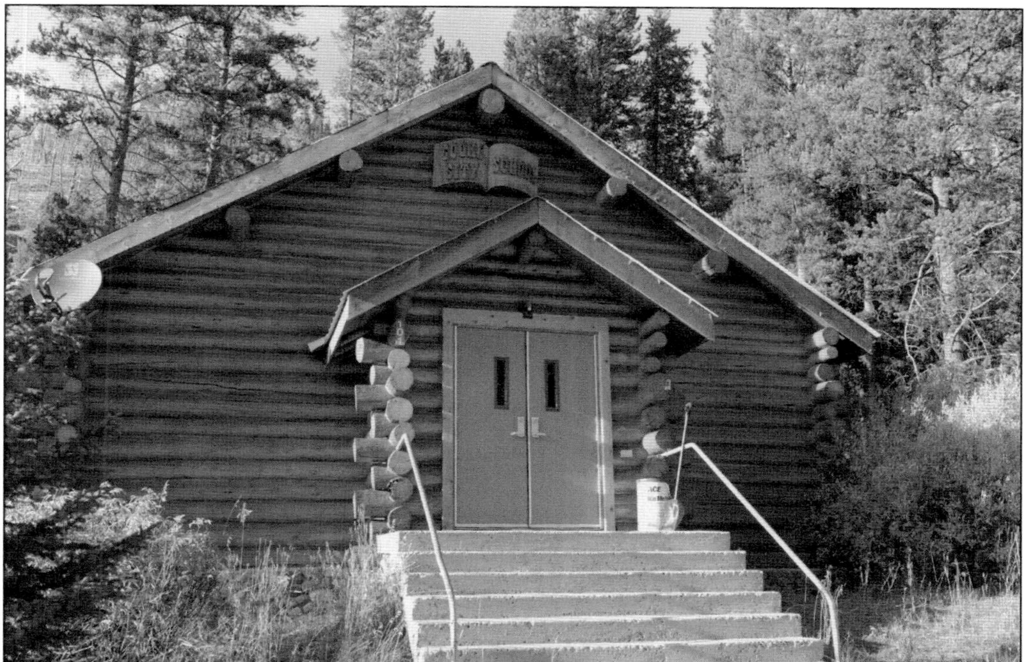

Cooke City's school is now located in the previous St. John's Catholic Church. High school students commute to school in Gardiner. A nondenominational church, Mount Republic Chapel, is located between Cooke City and Silver Gate.

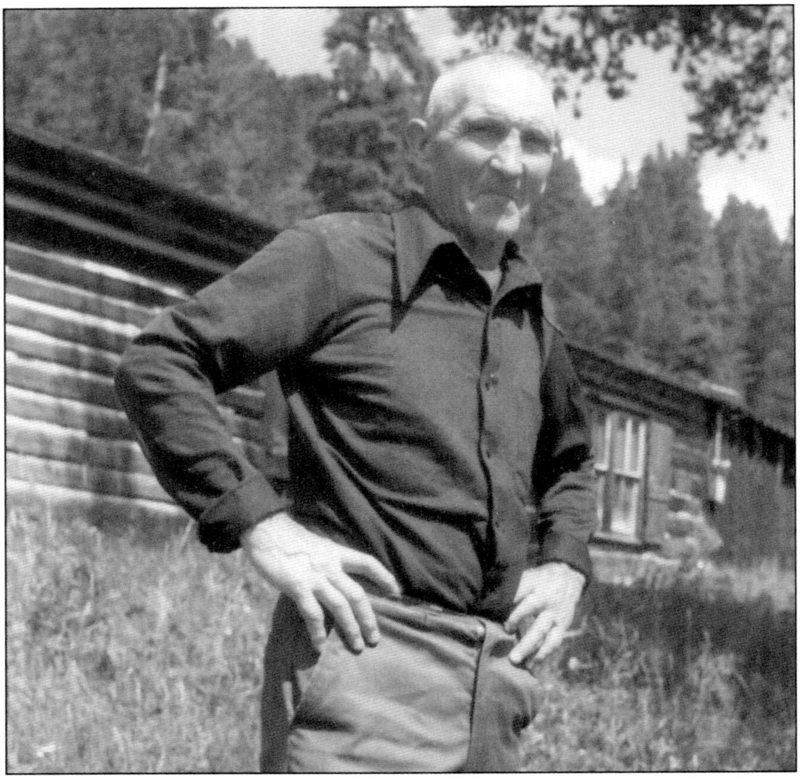

Tommy Garrison worked for Nick Freddenick at Buffalo Montana Mining Company. He was known as Cooke City's hermit and lived in a cabin on Lulu Pass that burned in the 1988 fires. With no place to stay, he lived in his MGB next to the Cooke City Store. When the cold temperatures came, concerned Cooke City residents cared for him until his death in 1989.

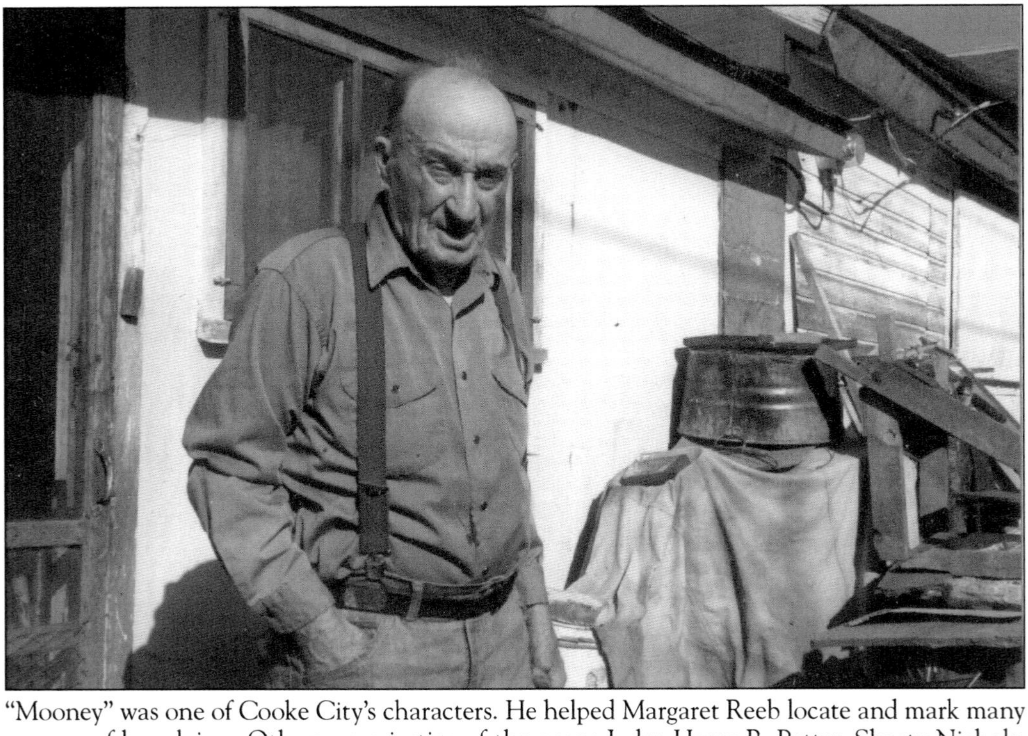

"Mooney" was one of Cooke City's characters. He helped Margaret Reeb locate and mark many corners of her claims. Others were justice of the peace Judge Harry B. Potter, Shorty Nichols, George Ogden, Dave Fleming, "Frenchy" Duret, and John Yancy, just to name a few.

Four

GOLDEN BROWN

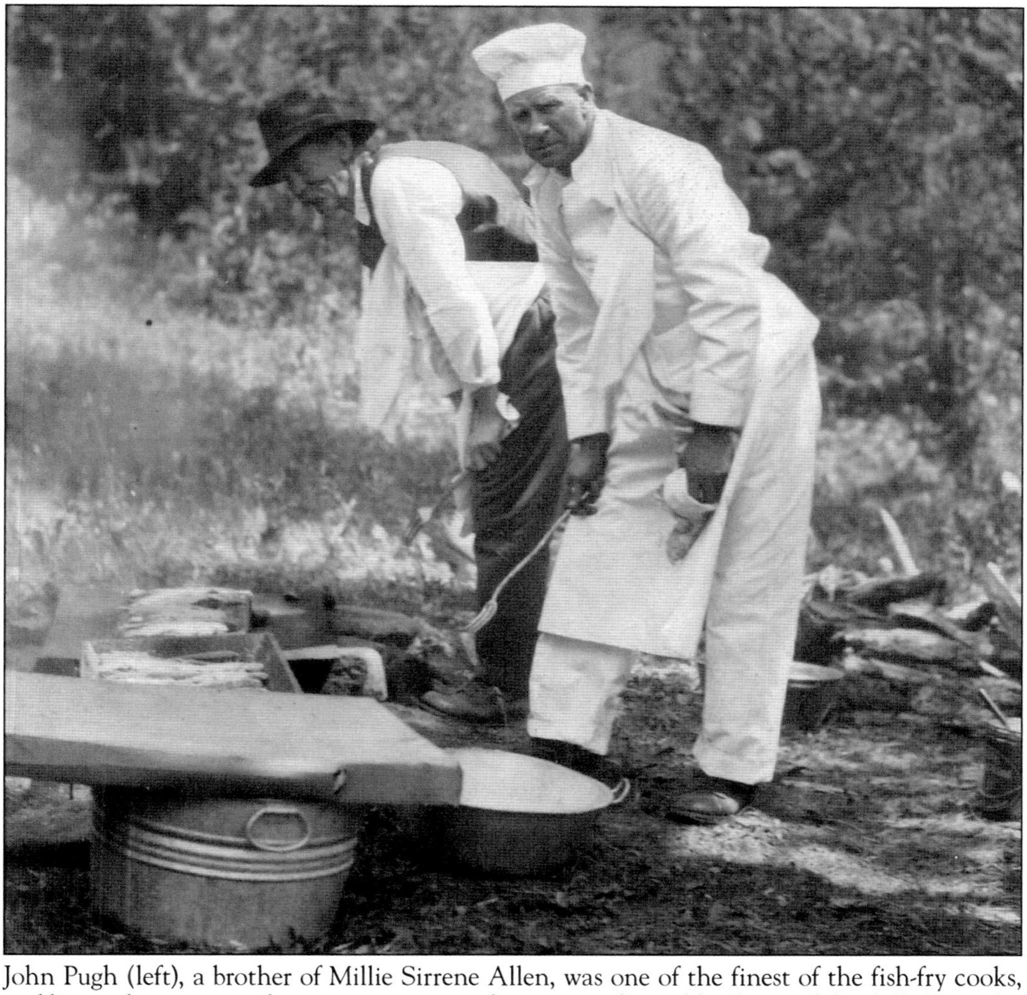

John Pugh (left), a brother of Millie Sirrene Allen, was one of the finest of the fish-fry cooks, and he was known over the years to turn out the most perfect golden brown fish of all those who took on the task Some rainbow and eastern brook trout were fried up by Pugh and John Burke (right) for the Old-Timers Fish Fry.

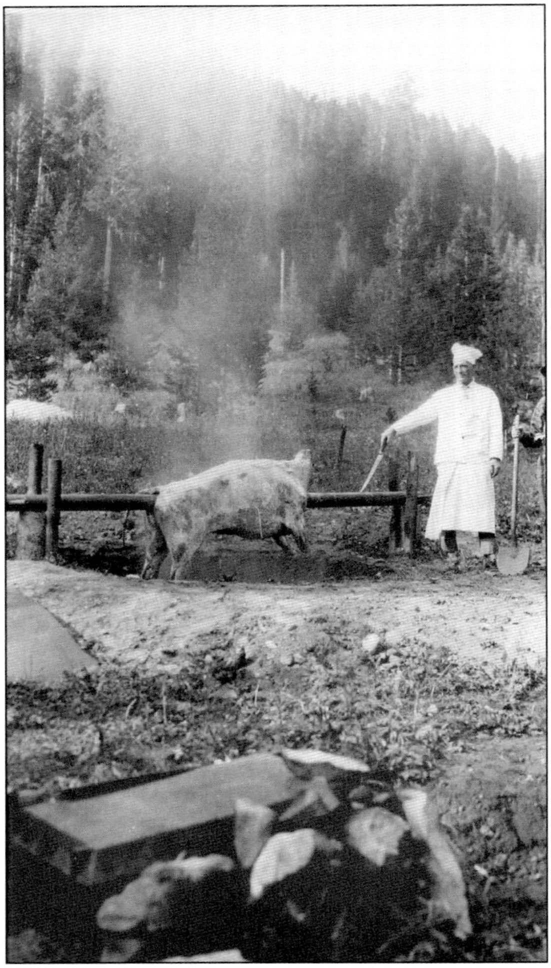

In 1925, over 1,000 people came to Cooke City to celebrate with the old-timers and meet their old friends who believed that the extensive mineralization of the district would be realized. Besides the hundreds and hundreds of fish being fried, a steer was roasted on a spit in anticipation of a large crowd.

Plenty of food was provided under the direction of Chester Shaw and aids at Shaw's Camps, the Cooke City Store, bars, restaurants, lodging hotels, and the work of local people.

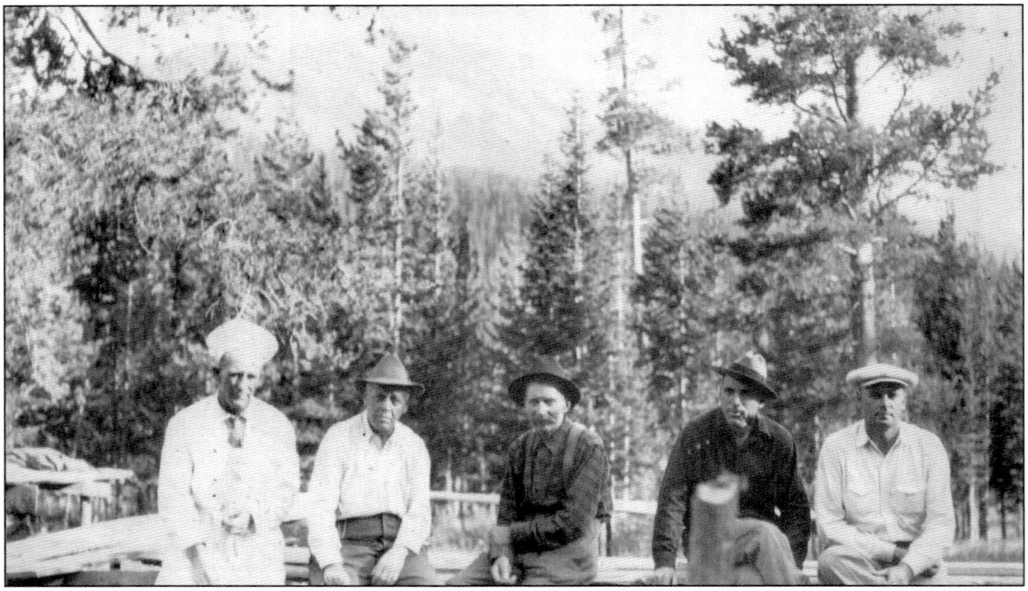

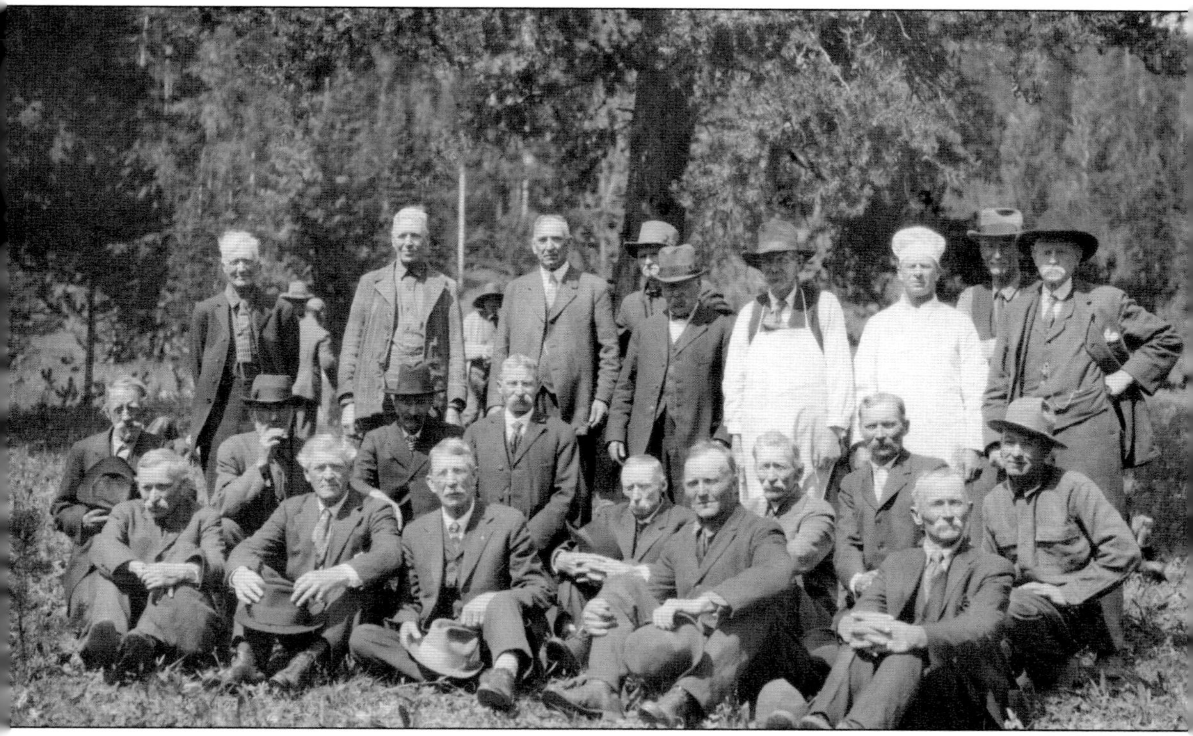

This photograph is of the old-timers at the annual fish fry. They include G.I "Trapper" Gibson, Jack Ornsby, Gus Solomonson, Nick Tredennick, Frank Mather, John Curl, Sam Mather, Martin Ranmael, Dan Ross, Bill Nichols, Bob Morton, Pete Branser, John Burke, John Pugh, Bob Kerr, and Pa Kearns, among others.

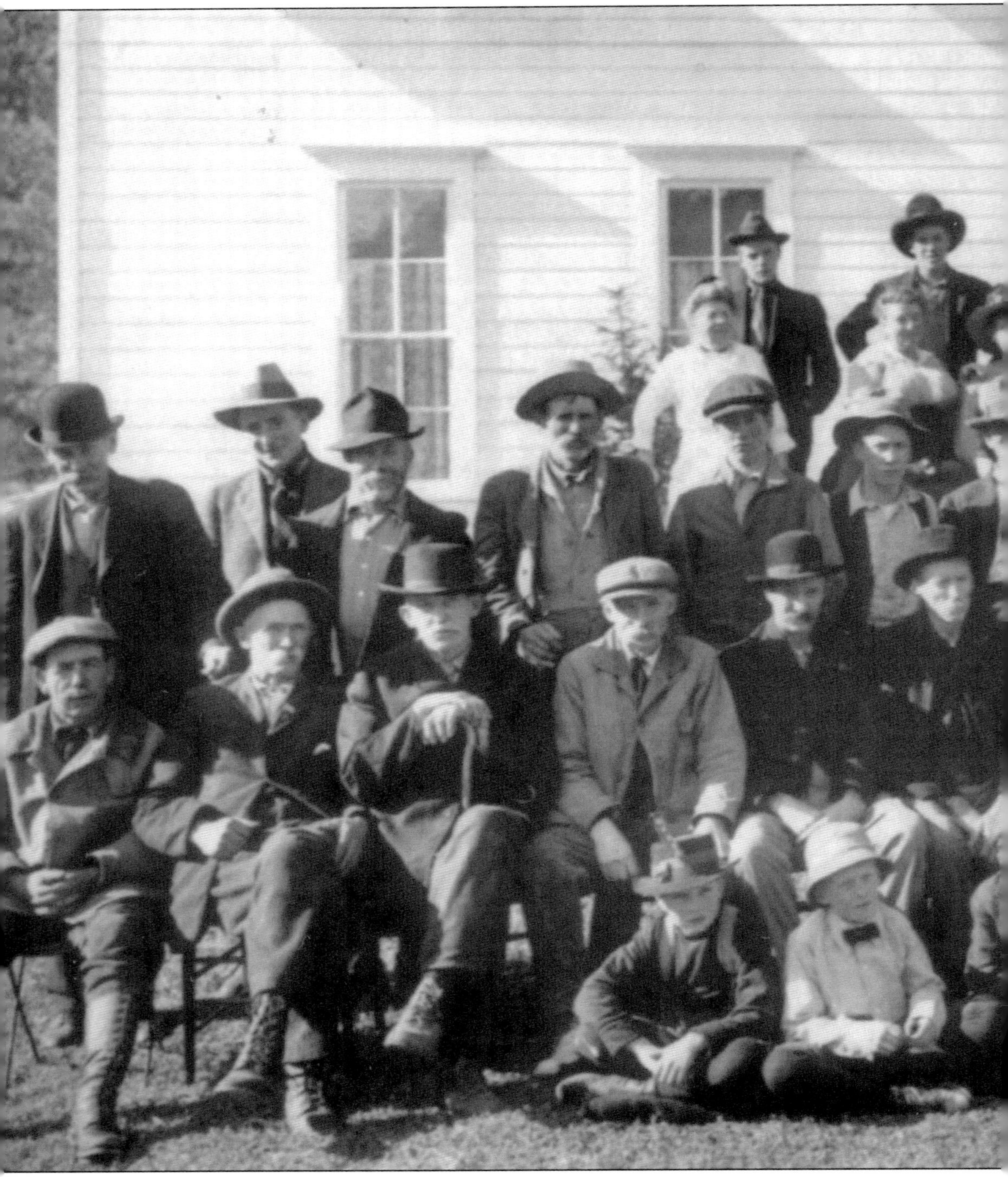

The McKay house on Main Street provided the backdrop for this fish fry photograph. The individuals are identified as (first row) Kenneth McDonald, Bobby Orr, Harold Soderholm, Raymond Holland, Gene Underwood (seated on a stool), Pauline Soderholm, Margaret Reeb, and Carolyn and Elizabeth Tredennick; (second row, from left to right) H. Reagan, Bill Nichols, Tom Harden, Jack Ormsby, Frank Mather, Sam Mather, Col. Dave Noble, Gus Soloman, Jack Allen, Dan Ross, Herman Eiff, Bob Mandeville, and Nick Tredennick; (third row, from left to right)

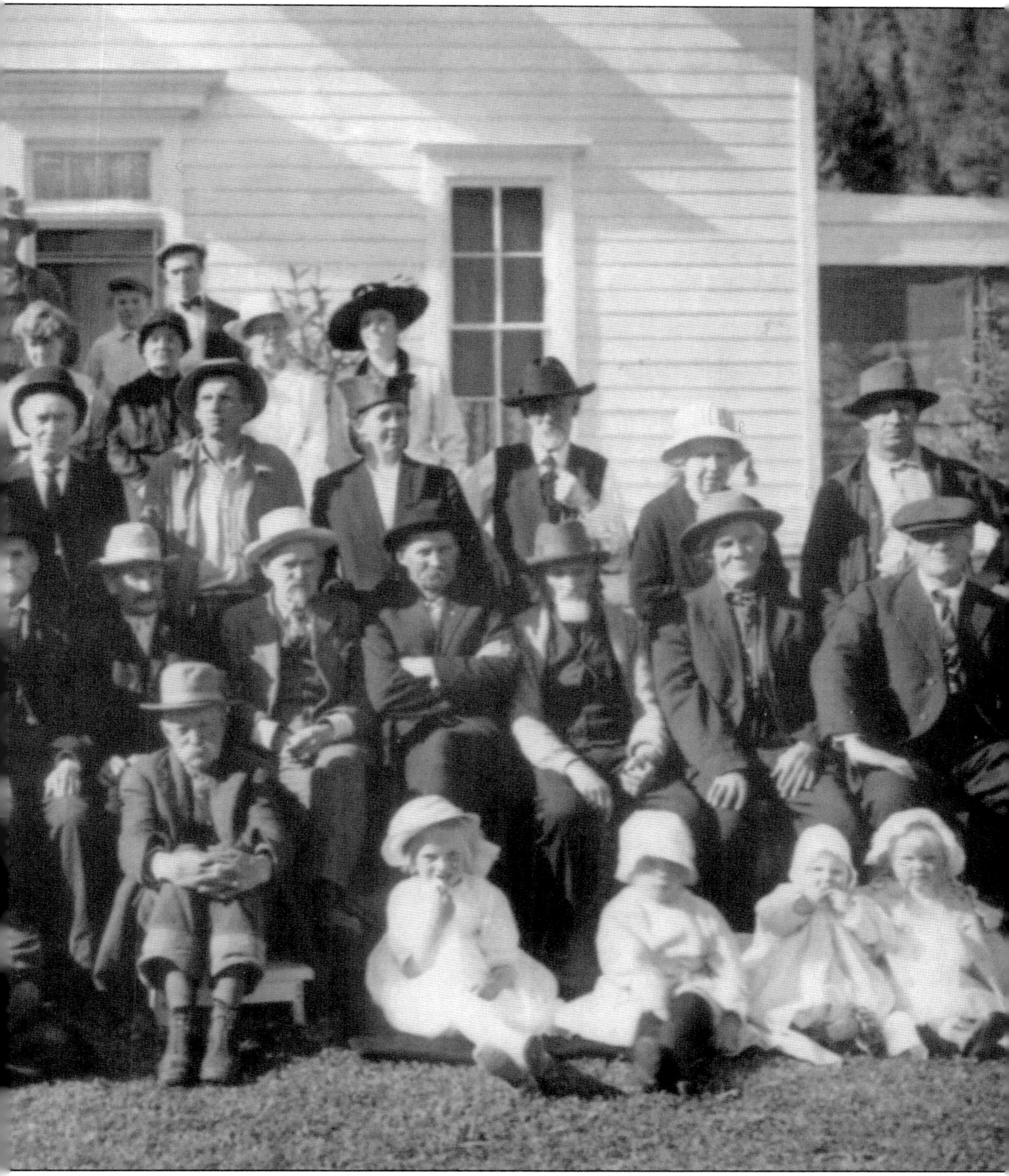

John Burke, Bill Tanzer, Jim Terrill, Dave Rhoads, Francis Brown, Randolph Holland, Cy Parker, Charlie Cannon, George Reeb, Mrs. George Reeb (Ingeborg), G.I. "Trapper" Gibson, Mrs. Bob Mandeville (Rebecca), and John Pugh; (fourth row, from left to right) Mrs. Noble, unidentified, Mrs. Jack Allen (Carolyn), Mrs. Robert McKay, Mrs. G.I. Gibson (Sarah), Mrs. John Harden (Hessie), and Mrs. Nick Tredennick; (fifth row, from left to right) unidentified, Clyde Gilbert (from Jardine, Montana), Robert I. McKay, and his two boys.

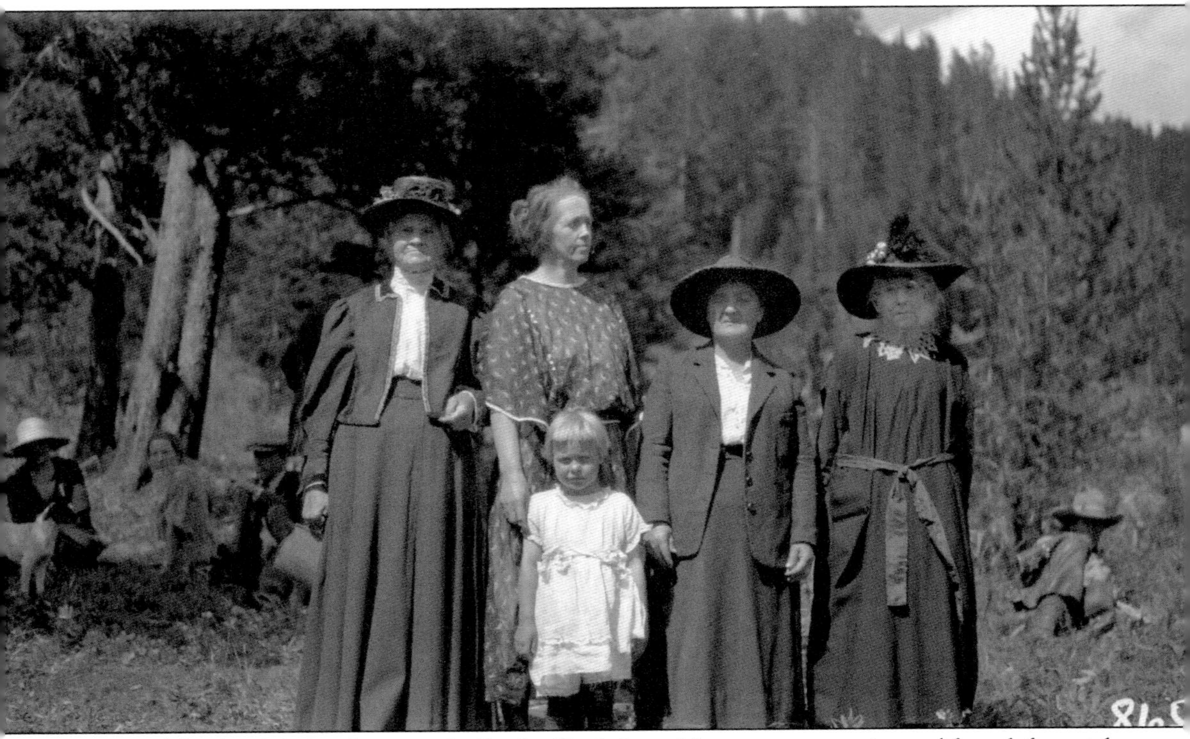

Posed for this fish-fry photograph, Cooke City's four oldest women are pictured from left to right as follows: Mrs. Anton Zukor (Anna), Ingeborg Reeb with her daughter, Georgia, "Good Eye" Gibson, and Rebecca Manderville. Mrs. Zukor and her husband lived in a cabin in the tall timber. For 25 years, the two drove down a tunnel into the side of the mountain. In his broken English, Anton said it was quite "hardt." They realized it was less work to dig a smaller tunnel, and Anton became stooped but kept pecking away at the face of the tunnel. After harrowing days with the Indians in the Joliet area, Ingeborg Holland (Reeb) and her husband, B.R. Holland, moved to Cooke City, where he worked his claim and did representing. (Courtesy of National Park Service, Yellowstone National Park, YELL 33264.)

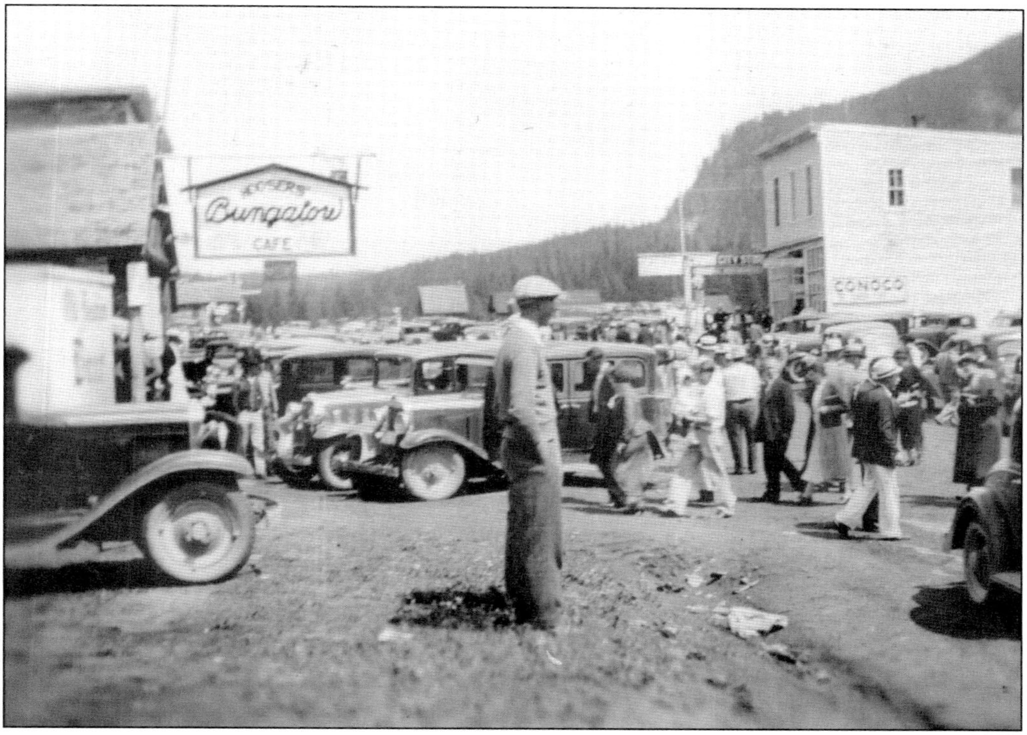

For this fish-fry celebration, 175 cars were counted driving into town. It was an event not to be missed and the last social event for the community before winter set in. Motion picture producers were on hand to record the event. (Courtesy of Nancy Adkins.)

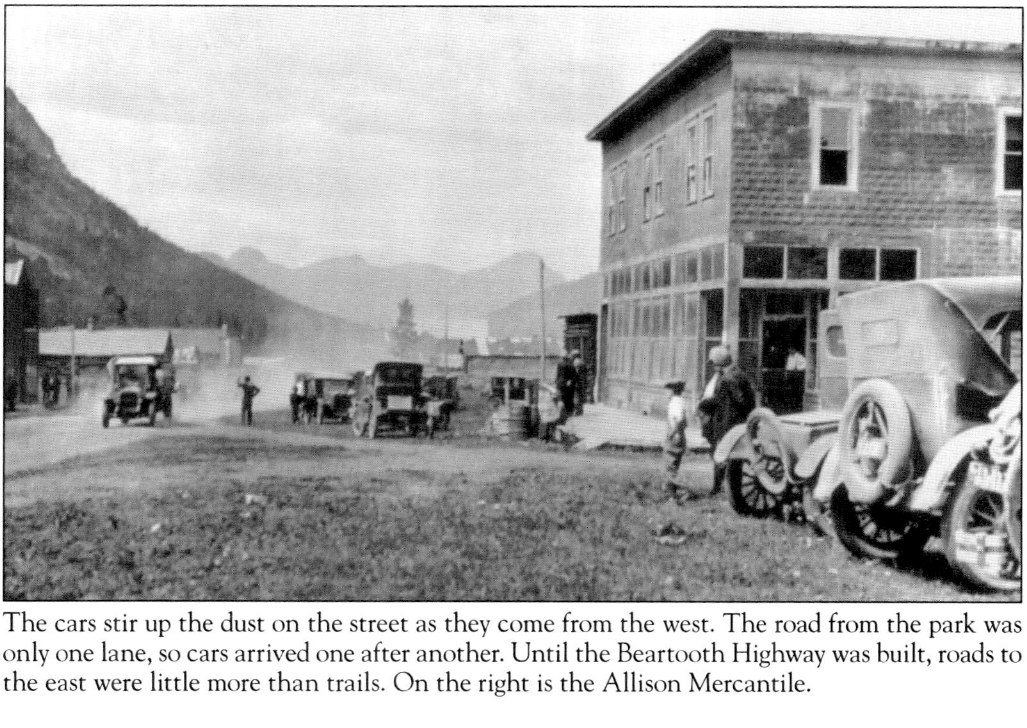

The cars stir up the dust on the street as they come from the west. The road from the park was only one lane, so cars arrived one after another. Until the Beartooth Highway was built, roads to the east were little more than trails. On the right is the Allison Mercantile.

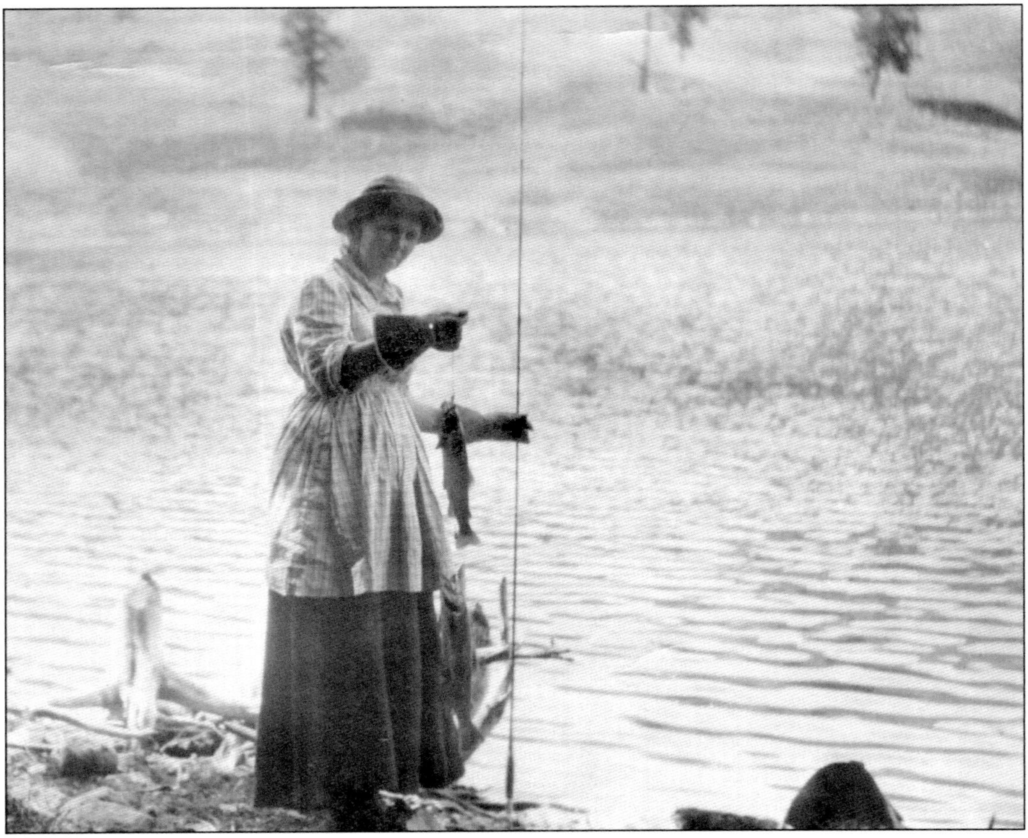

This is a nice catch for this woman.

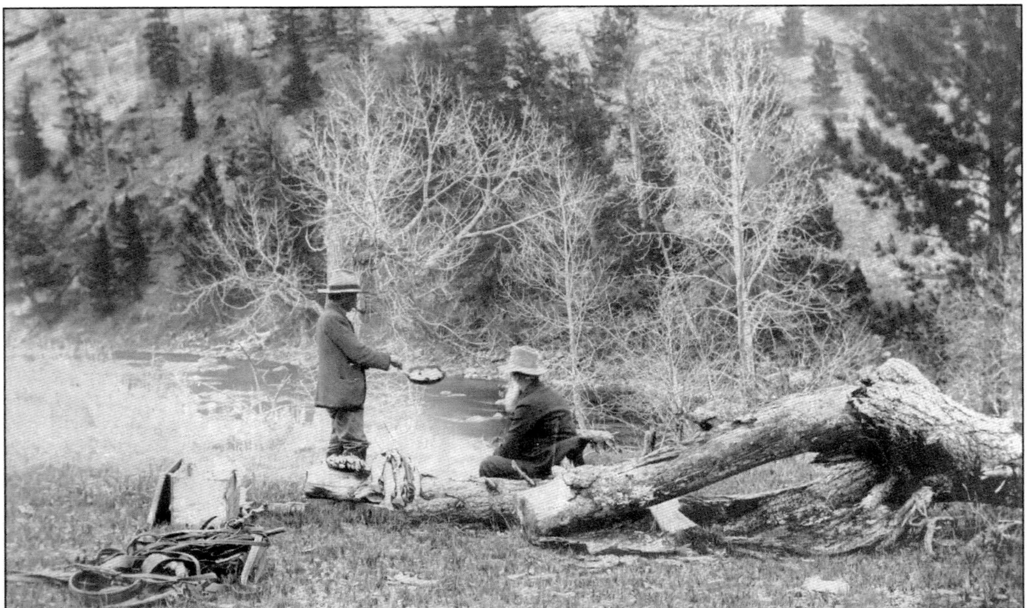

Lakes and streams abound with fish, including rainbow, brown, cutthroat, and brook trout. These gentlemen are about to sit down to a fine meal of fresh fried fish.

Five

SNOWBOUND

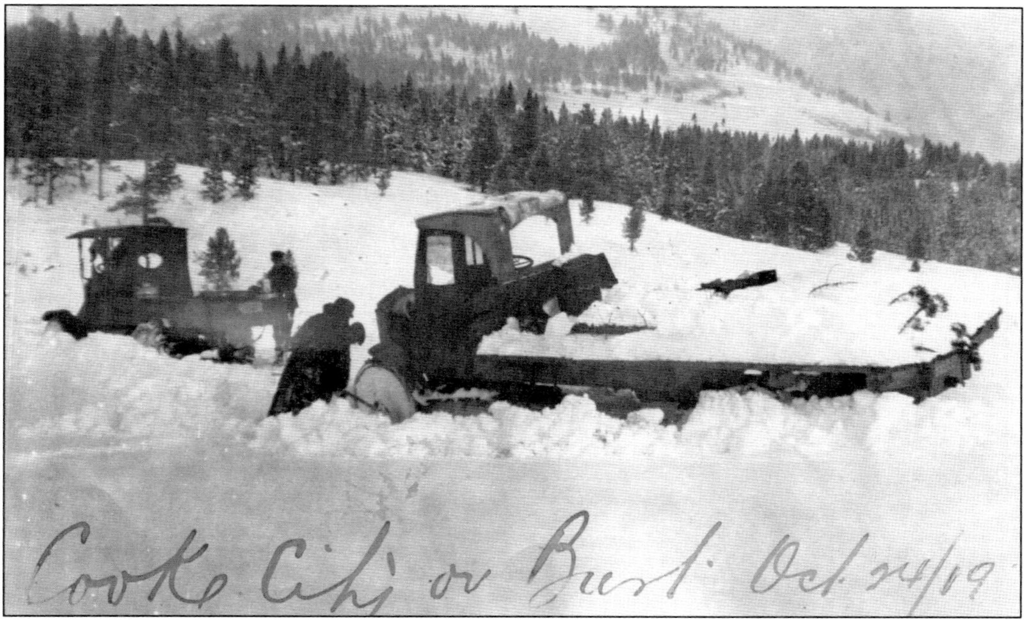

This image is captioned, "Cooke City or bust Oct. 24/19." At times, Cooke City was almost completely isolated from the outside world. In 1888, David Dobson carried the mail all winter without missing a service. He had snowshoed over 2,000 miles, been snow blind and lame in one leg, had his nose frozen, and had lost overshoes and gloves repeatedly, but in spite of all, he never failed to deliver the mail.

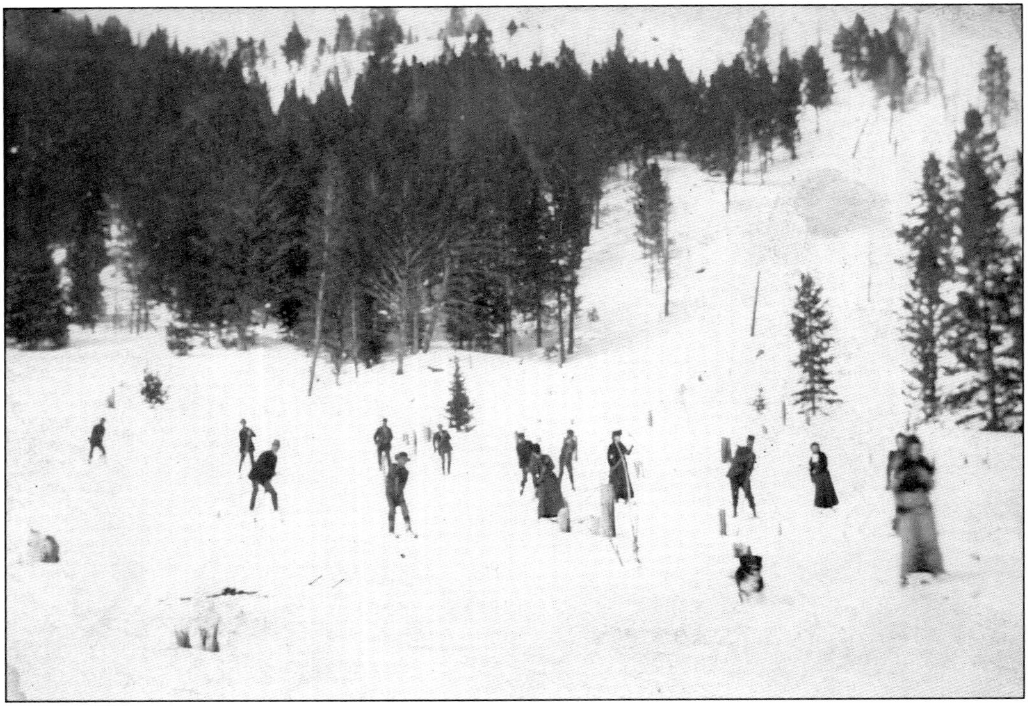

Even with the deep-freezing conditions winters often brought to the area, there was no lack of fun in the community. Young and old participated in skiing and sledding parties. The women met for hooked rug bees and to play bridge. While the children slept, the adults danced at a chosen saloon. There was plenty of coffee and food.

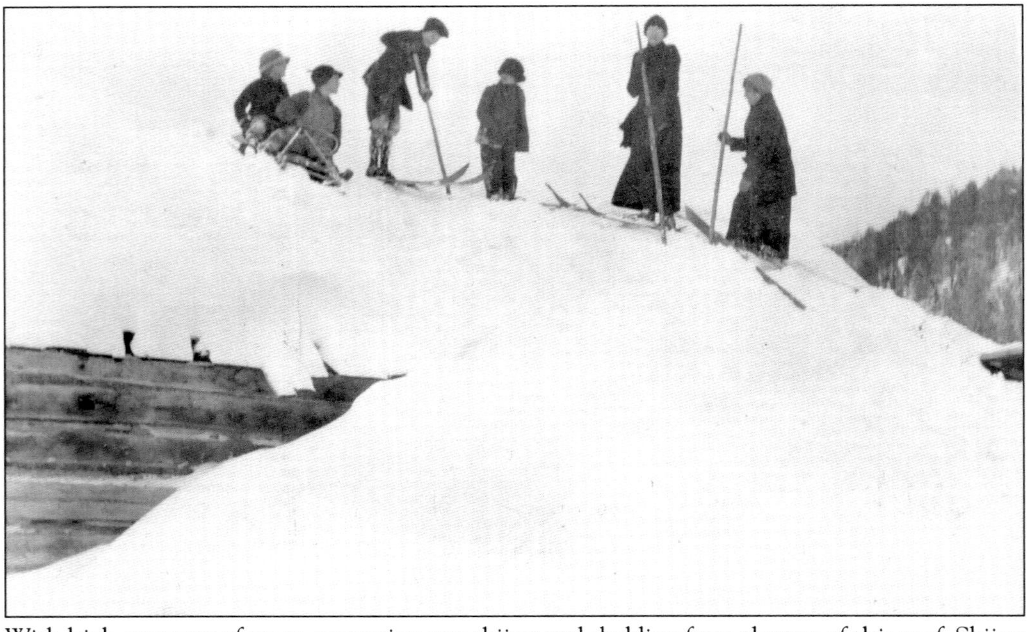

With high amounts of snow, recreation was skiing and sledding from the top of this roof. Skiing was the best way to get around town at times.

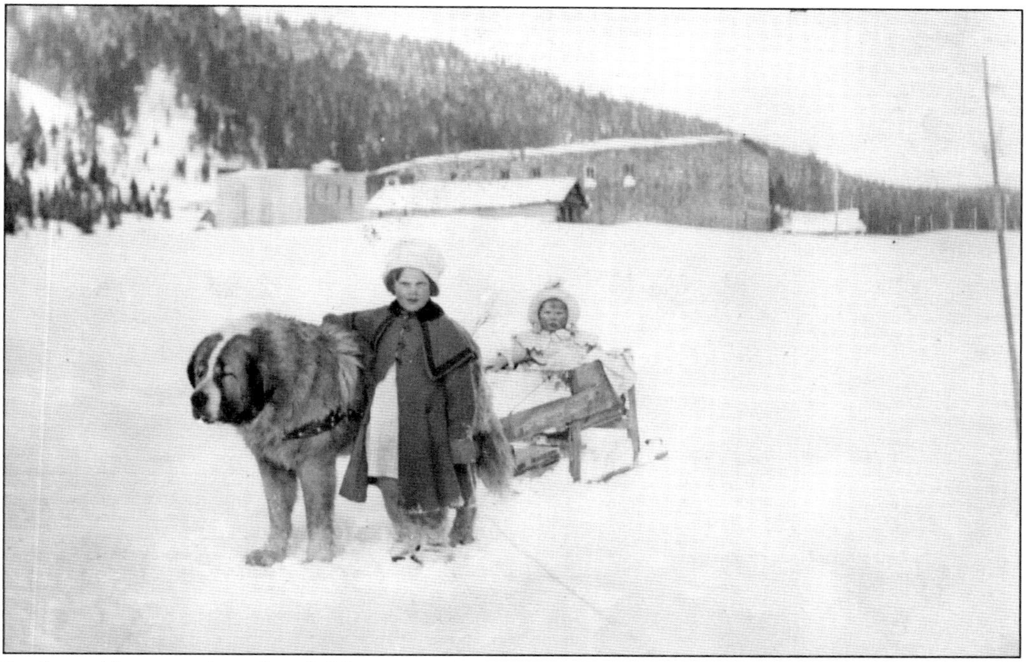

In the coldest weather, the children played outdoors. These two youngsters have a dogsled. In 1886–1887, when supplies ran low, E.E. Van Dyke took a dog and sled to Gardiner, which was 120 miles away. He sold everything on the sled for $1 a pound, a small price for his hardships during those three days it took.

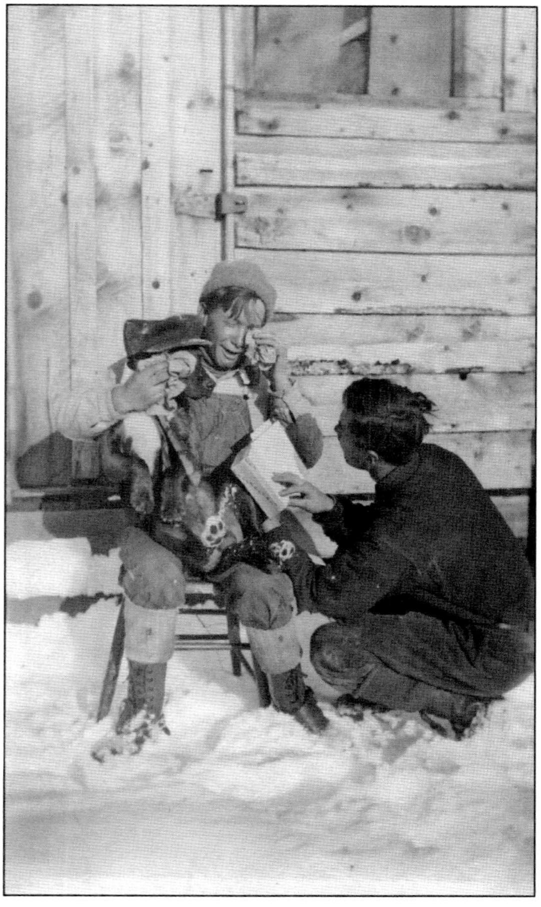

During the winter, residents had to have something to lighten the short days. Rand Holland (left) is clowning around trying to dress his dog Tige. Ing Gronvold appears to be giving him advice.

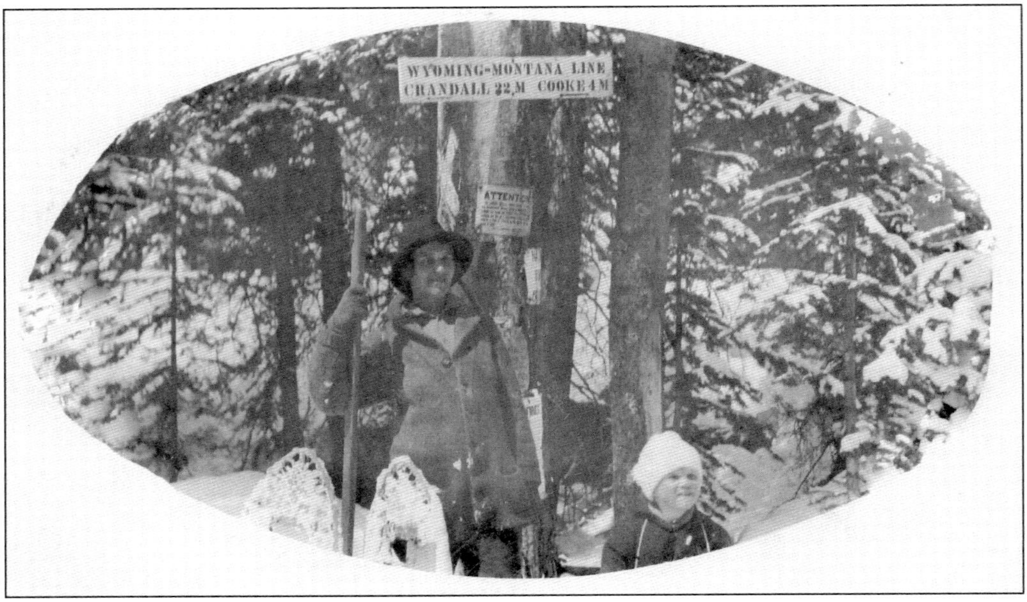

The picture on this postcard is on the Wyoming-Montana line. Crandall is 22 miles away, and Cooke City is 4 miles from here. At one time, Cooke City was in a no-man's land, as neither Wyoming nor Montana had it mapped.

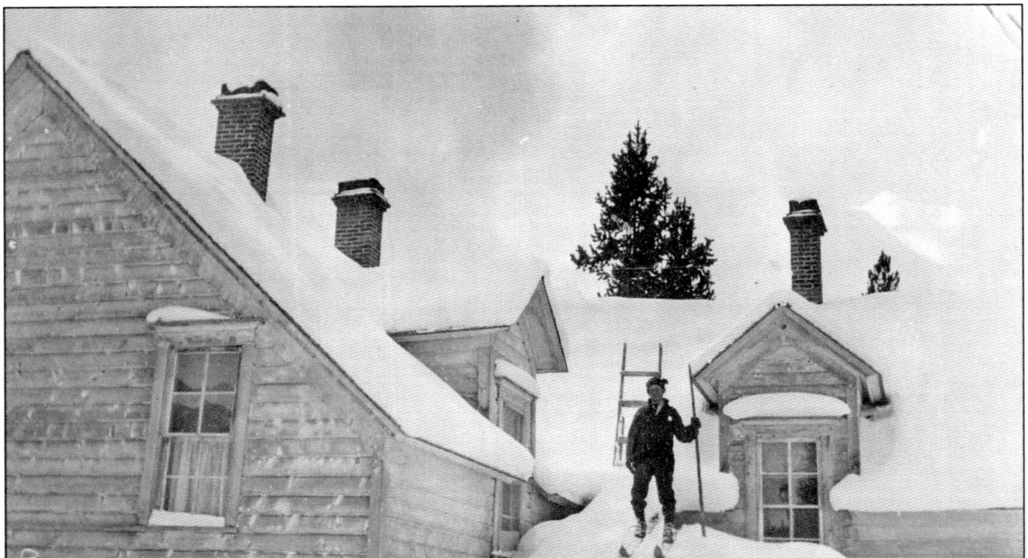

Roofs had to be shoveled when they were laden with snow. Skiing off the roof of the McKay house was an extra long run.

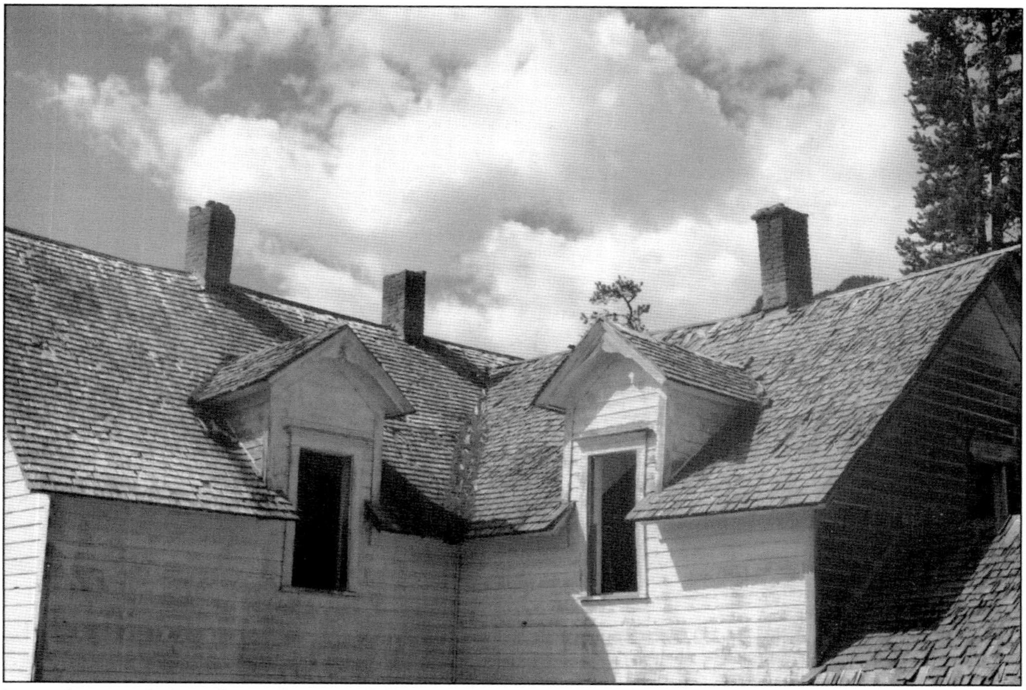

Paint has peeled and windows and doors are gone, but the McKay house still stands today. It is often referred to as the "White House."

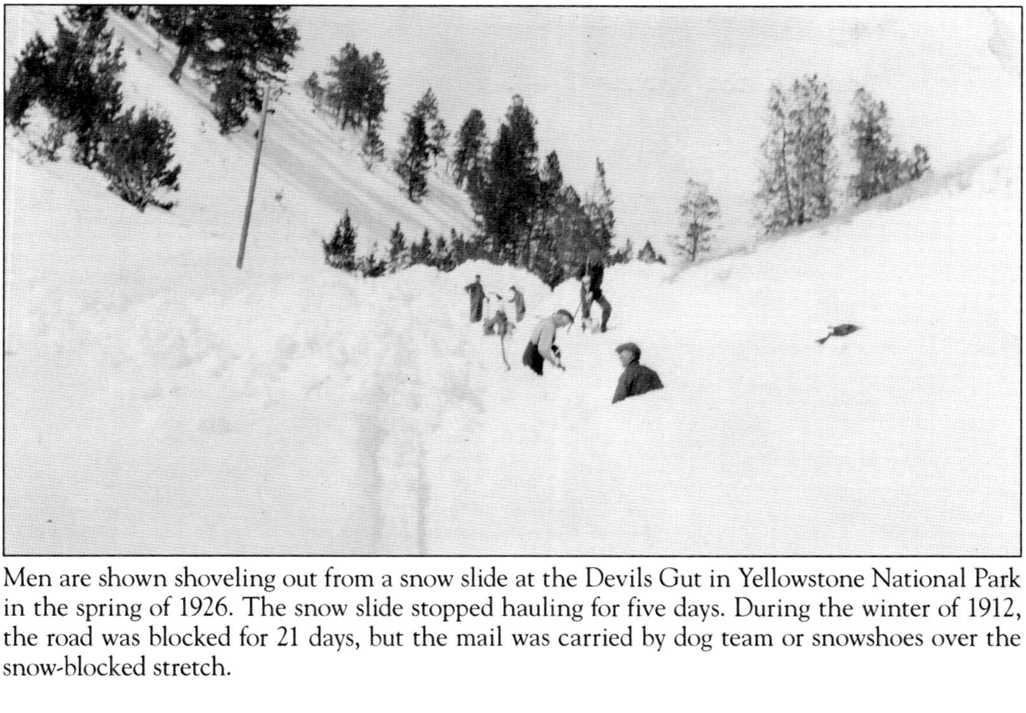

Men are shown shoveling out from a snow slide at the Devils Gut in Yellowstone National Park in the spring of 1926. The snow slide stopped hauling for five days. During the winter of 1912, the road was blocked for 21 days, but the mail was carried by dog team or snowshoes over the snow-blocked stretch.

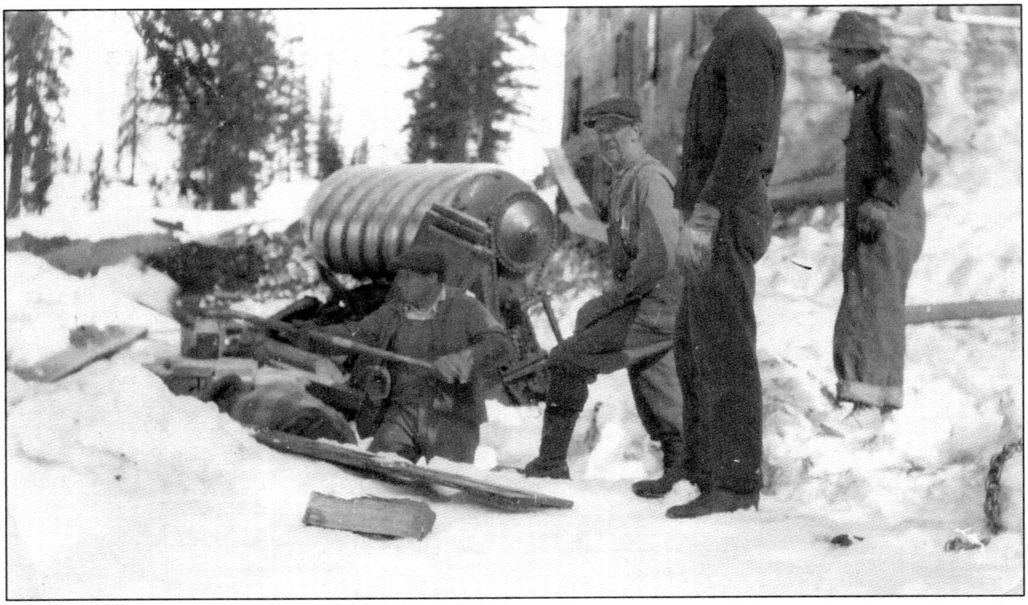

When the road gave away, the sled broke, upsetting the snow motor. Men worked to pull out the motor. John Pugh is on the left. The snow was 6 to 10 feet deep near the sawmill.

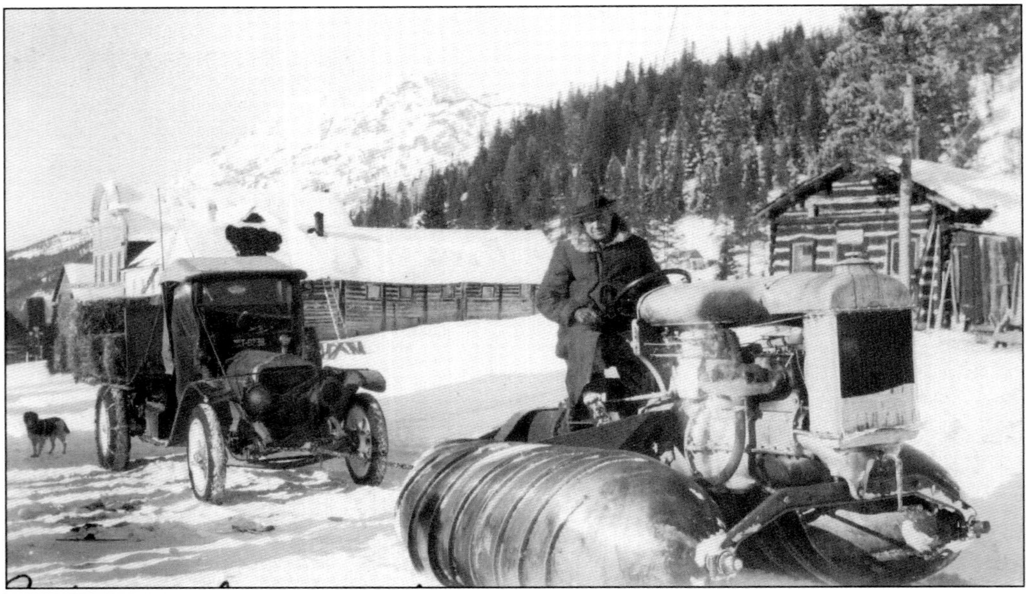

In December 1926, the snow motor pulled a load of hay into town. The snow motor rode on a snow road from four to five feet deep. Harry Stinson's cabin is in the background of the photograph. Harry was the justice of the peace and is known for keeping many historic records of Cooke City.

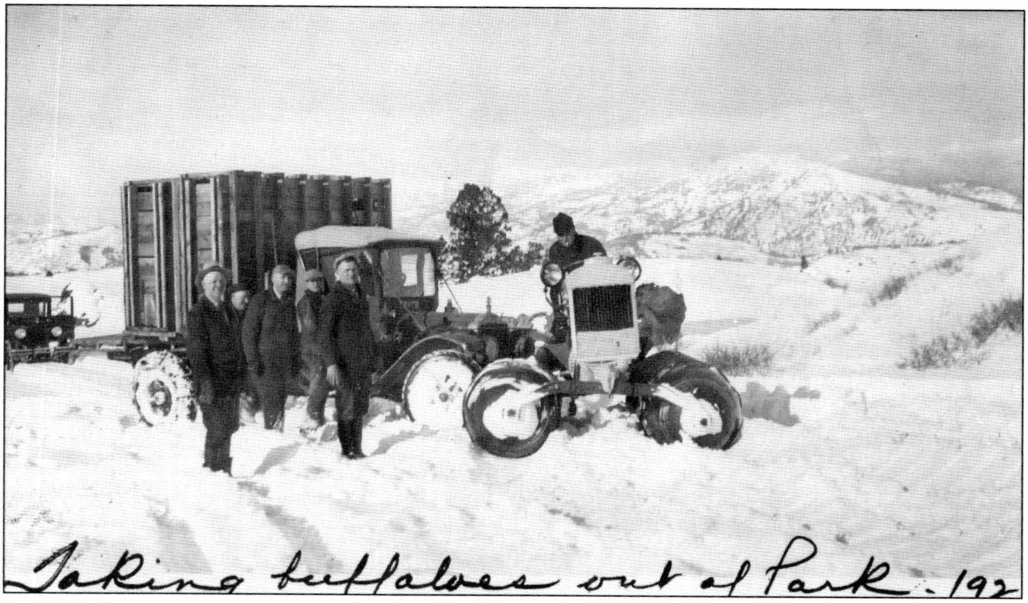

In 1926, the snow motor was used to haul buffalo out of Yellowstone. After every snowfall, it was used to break a trail from Cooke City to the camp, about seven and a half miles. When the snow was too deep, the plow did not operate efficiently.

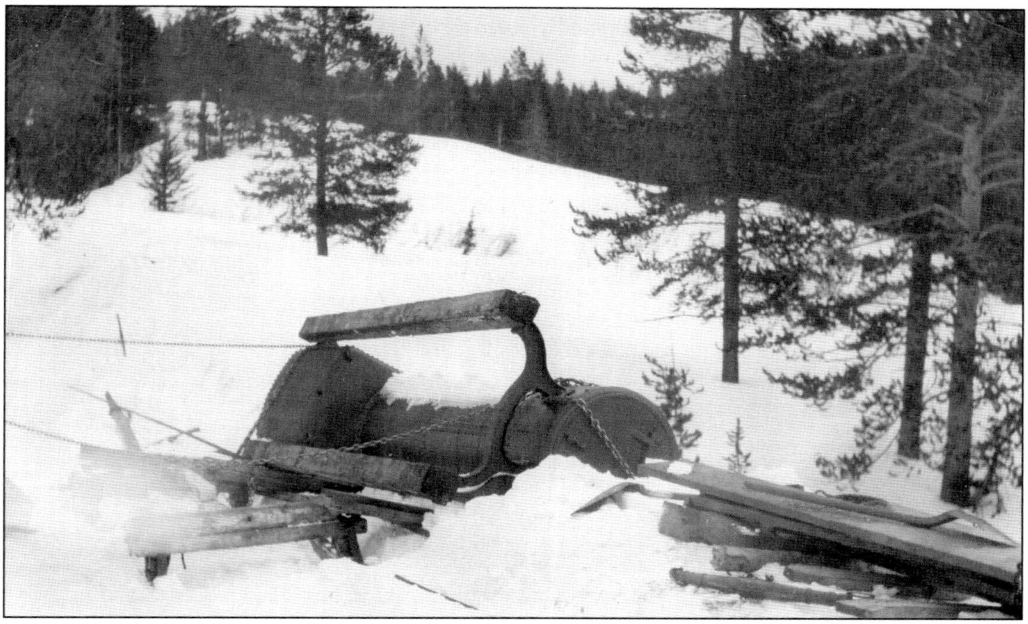
With the help of horses and a snow motor, it took two days to pull this boiler out of the snow and haul it to camp.

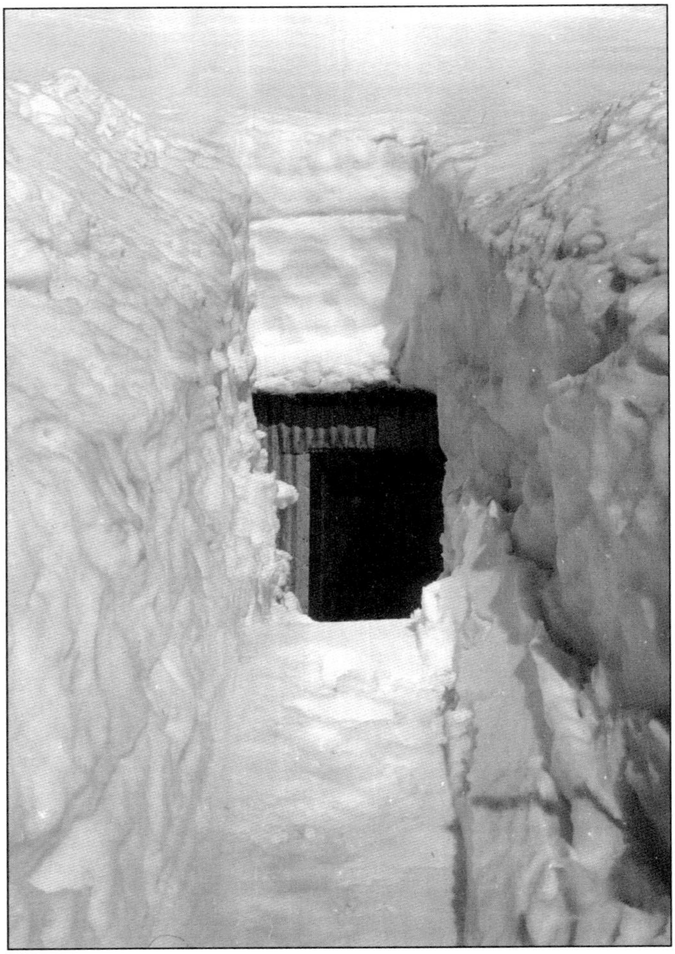

Snow and cold were especially bad in 1886–1887, with snow 10 to 12 feet deep. On New Year's Day in 1886, a small slide buried the entrances of two tunnels on Lulu Pass. Snow was shoveled to help clear the entrances. Later in the day, a few men returned to finish, and Nick Tredennick was a witness as the whole mountain of snow broke loose and swept away everything before it, including Clarence Martin and Tony Wise. Their bodies were not found until summer.

Snow became so deep it was necessary to cut steps on an incline. Tunnels had to be dug from one building to another, and windows had a shaft to provide light. Candles had to be lit, and kerosene lamps had to burn day and night.

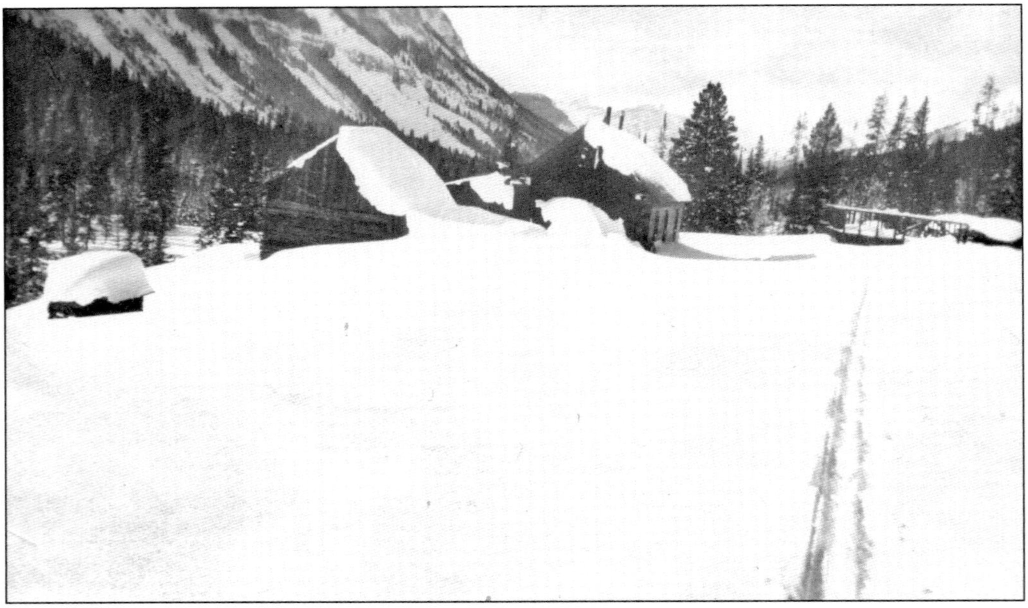

The snow motor enabled cars to be driven through tunnels in the deep snow. However, they were limited to one-way traffic.

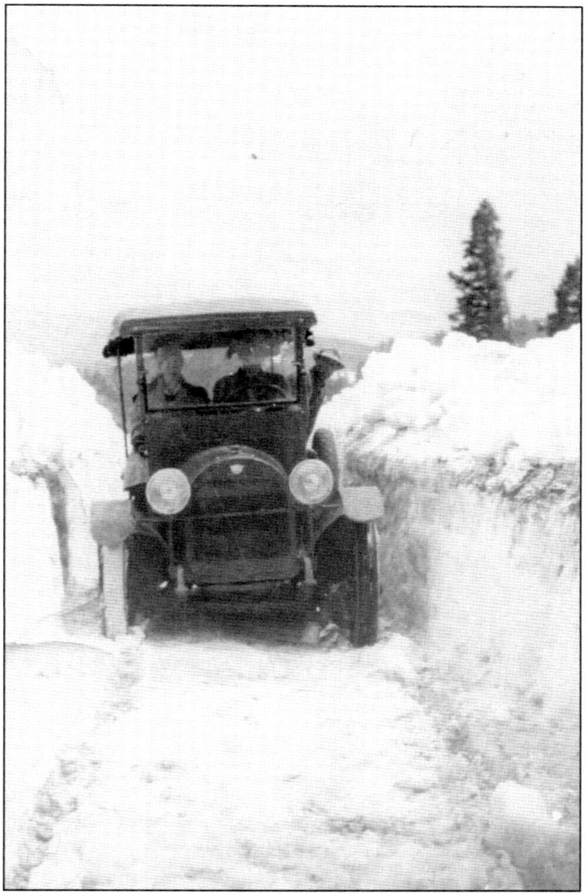

Ing Gronvold and Harry McDonald are ice fishing on one of the mountain lakes. Cooke City was remote in the winter, and when food ran low, fresh fish was welcomed. It was written that to keep from starving, poached elk was hung behind a building's false wall to provide meat to some hardy souls who stayed.

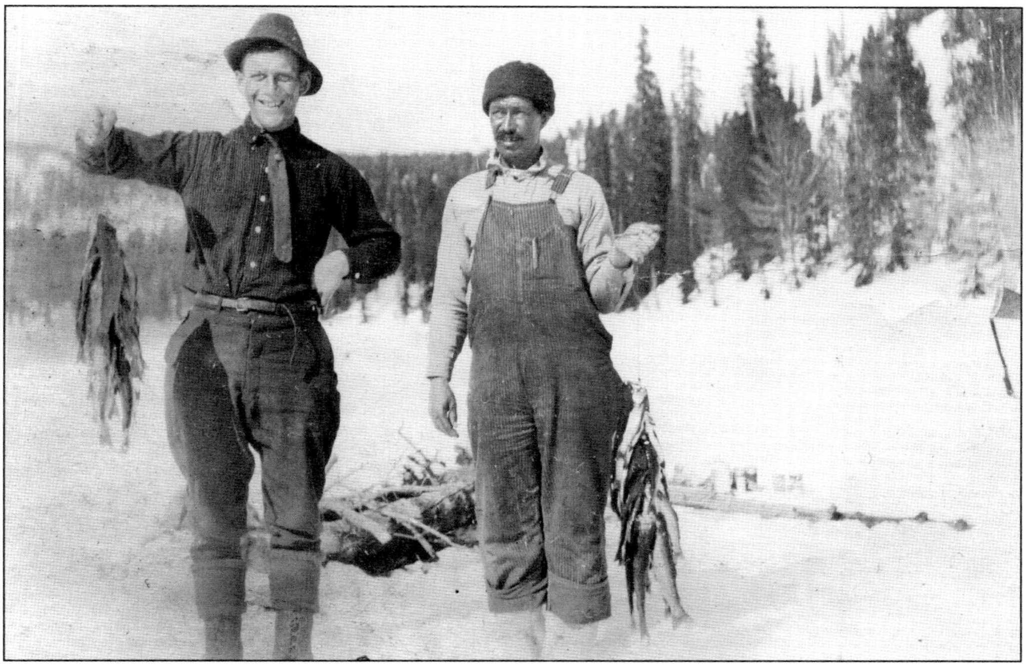

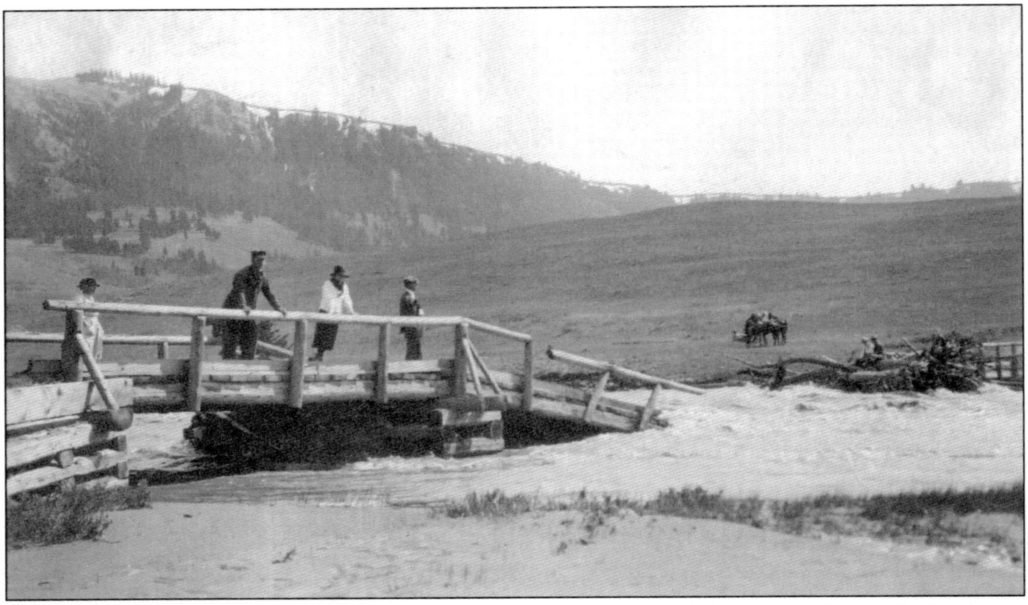

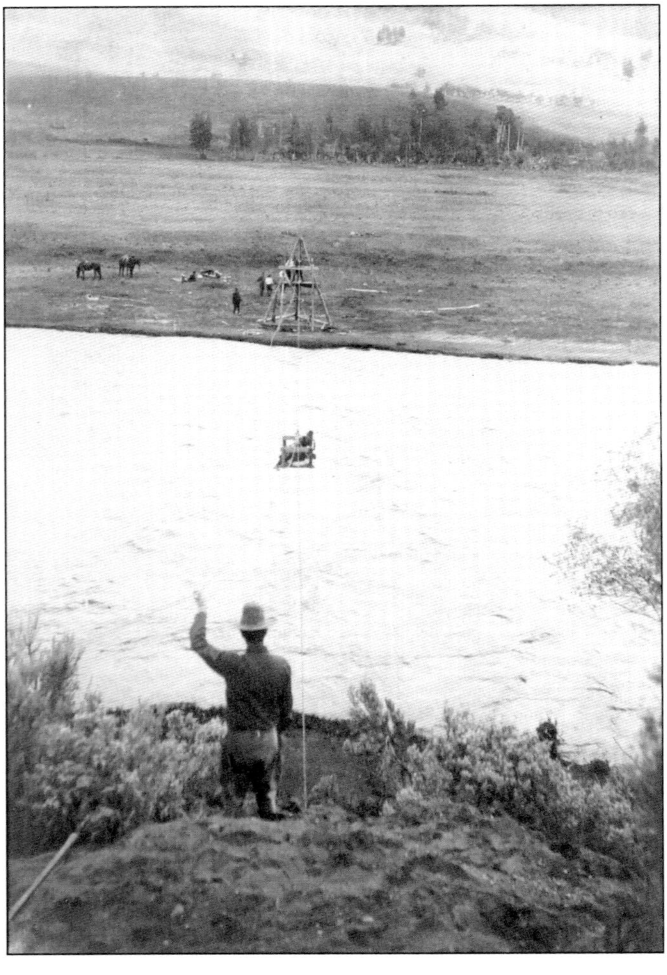

This image shows the bridge across the Lamar River as it was being washed out by spring flooding. Residents ate a lot of oatmeal when supplies were exhausted.

This aerial tram crossing the Lamar River until the bridge could be replaced was installed by G.L. Tanzer on June 14, 1918. It enabled supplies to be brought in and people to come and go. The tram was highly approved by Maj. George E. Verrill of the Yellowstone National Park Engineering Corps.

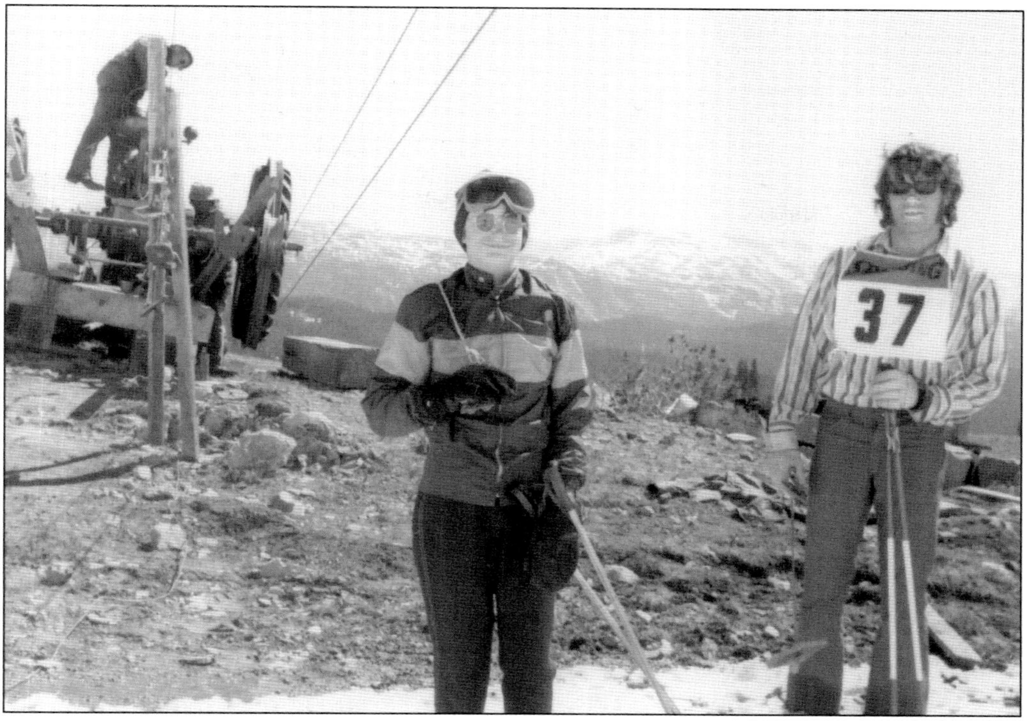

With snow remaining in the mountains, ski school could be held in June on Henderson Mountain.

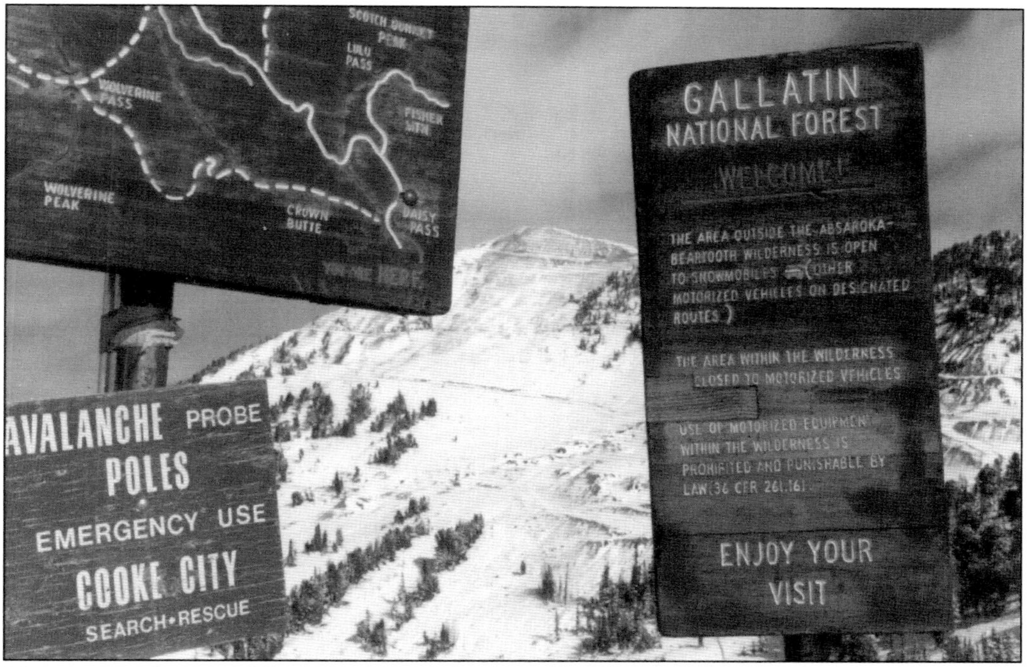

This sign at the top of Daisy Pass is a reminder that the area is Beartooth Wilderness with restrictions and warns of avalanches. Steep mountain slopes with 25 to 40 feet of snow create situations where avalanches occur. Snowmobiling in the area can be dangerous, and lives have been lost. (Courtesy of Warren McGee.)

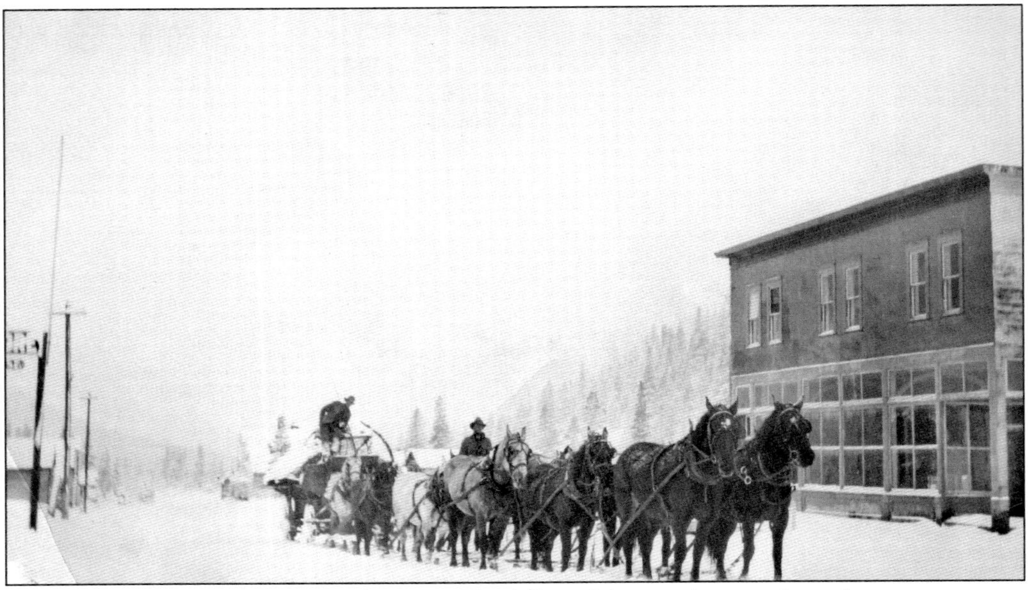

Freight is being brought into Cooke City. The Allison Mercantile is on the right.

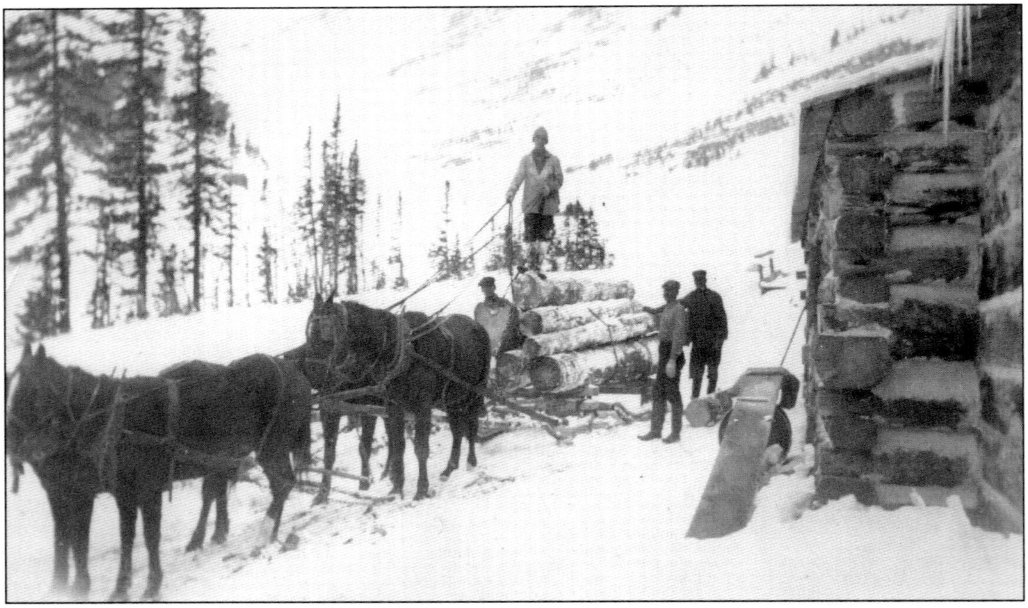

Logs were hauled year round. At the high elevation, trees grow slowly. The diameter of the logs demonstrate how old these were.

Six

Western Smelting & Power Company

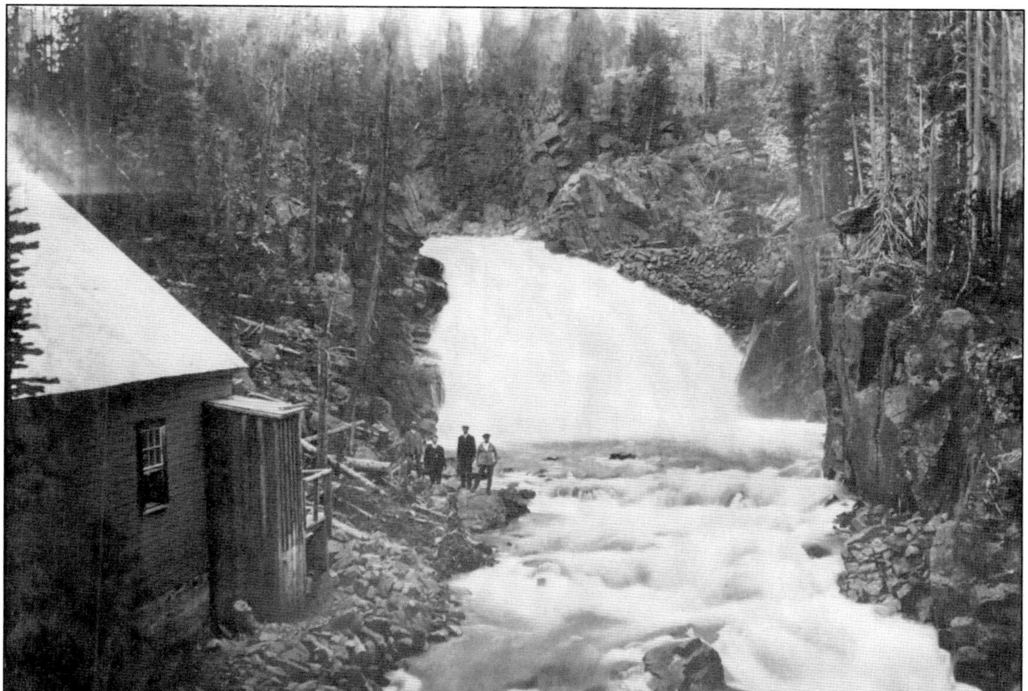

Pictured in 1914 and located four miles east of Cooke City and one mile from the Wyoming state line is the Big Fall on the Clark's Fork River with a fall of nearly 53 feet. The power plant on the left generated its power from this waterfall. The US government cooperated with Western Smelting & Power in the venture, granting a 50-year permit with the privileges to water rights of the Big Fall.

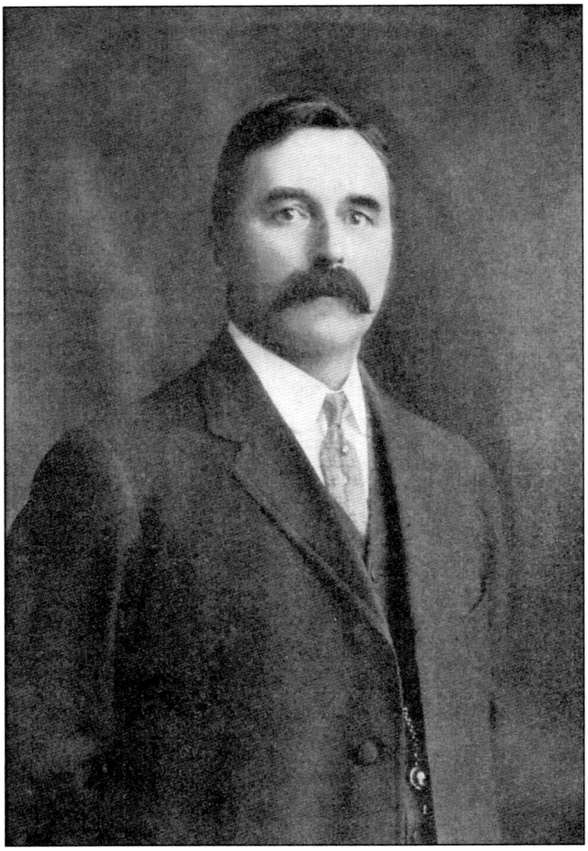

Dr. Gottwerth Lebrecht Tanzer, of Seattle, Washington, was president and general manager of the company that became known as the Western Smelting & Power Company. The magnitude of the era of extensive development brought new hope to Cooke City. New technology brought a solution to longtime problems of reduction. It had the most complete equipment ever assembled in the West. Because Cooke City was promising, Congress rendered assistance in its interests by building an easy-grade road from Tower Falls to the park boundary near Cooke City. Expenses made express wisdom and foresight. It was believed that the company would be in a position to pay sizable dividends to all shareholders.

This is the area before development of the power plant.

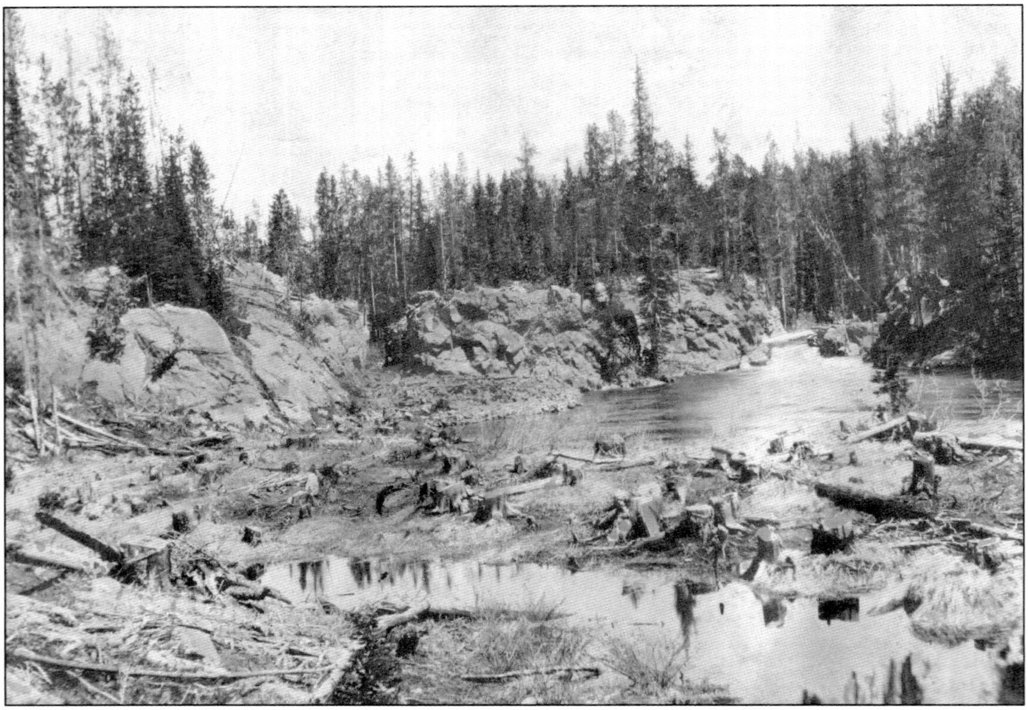

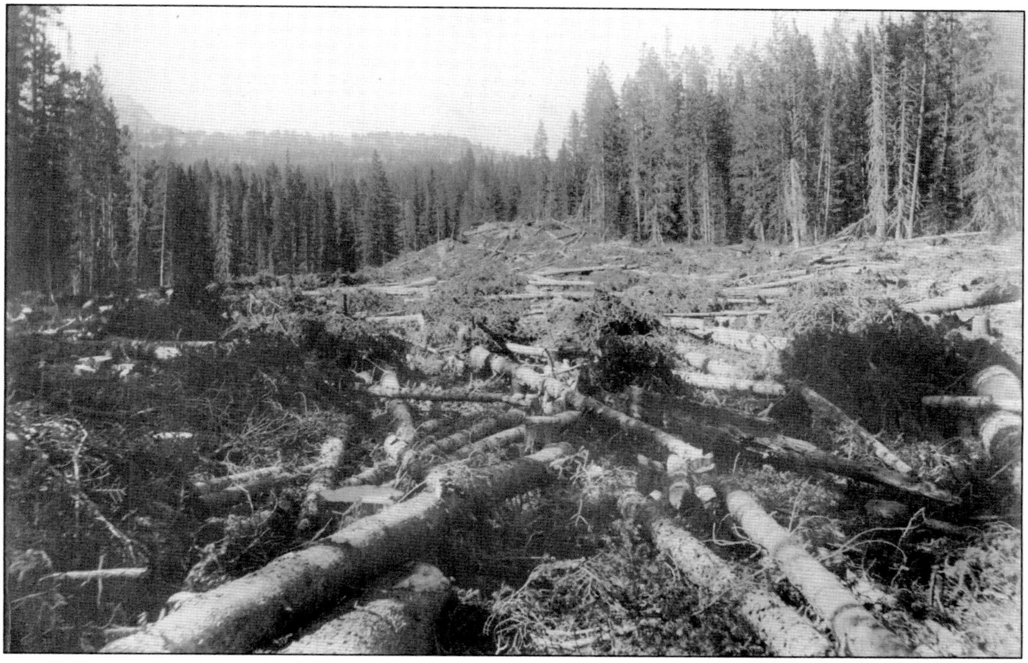

Near Soda Butte Creek, a force of 200 men cleared a path 200 feet wide and 3.5 miles long through dense forest for power transmission lines. The lines carried electricity to the different parts of the smelting plant, and 10 miles of telephone lines connected the smelter office with the hydroelectric plant, sawmill, mine plants, and Cooke City.

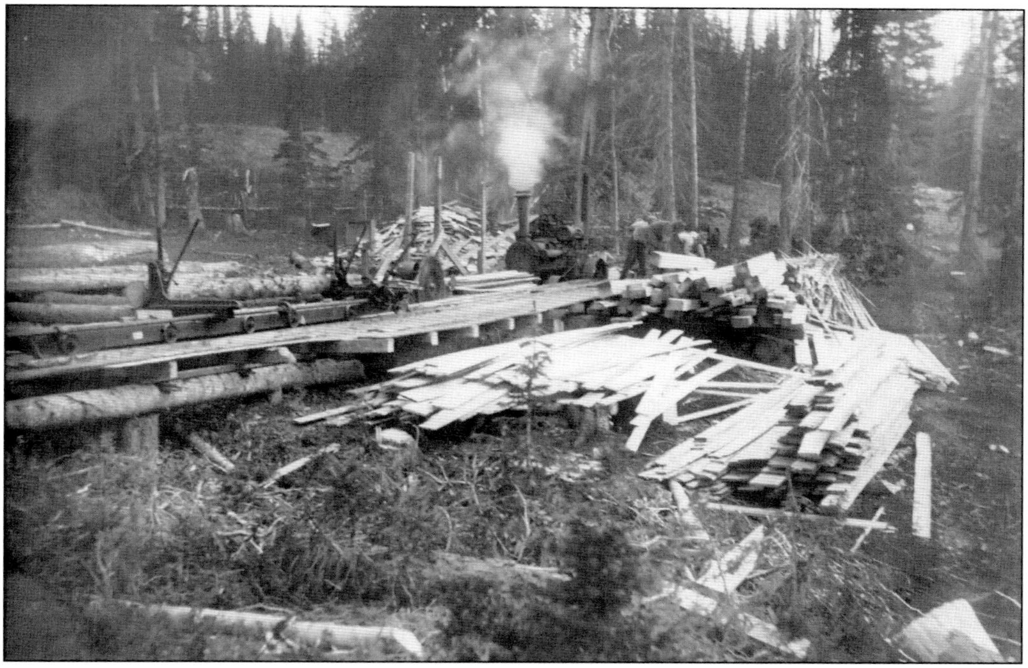

Timber and water are essential to mining. In 1918, this portable sawmill was set up about a mile from the smelter and sawed timber into lumber on the power transmission right-of-way. The government granted a permit for cutting timber and operating sawmills and for the use of land for power transmission.

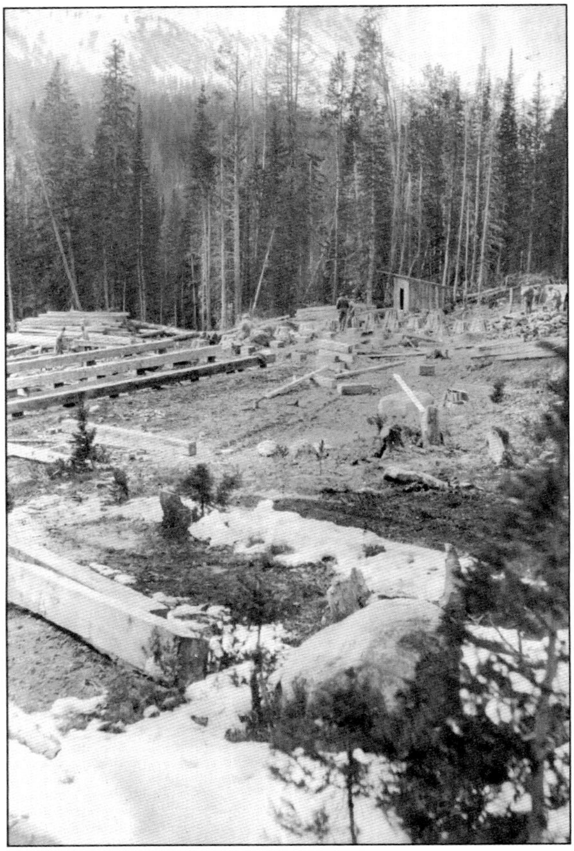

Excavation had to be done before foundations were laid for the flume and other buildings.

Tons of rock had to be blasted for the flume. The flume had double iron headgate hoists.

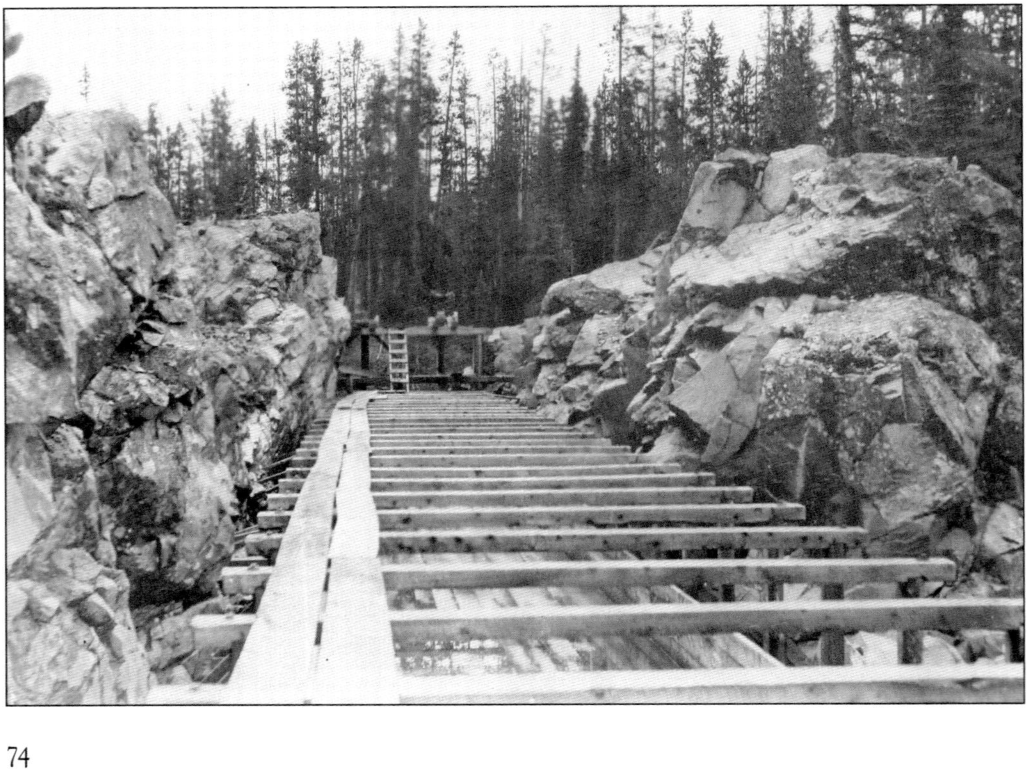

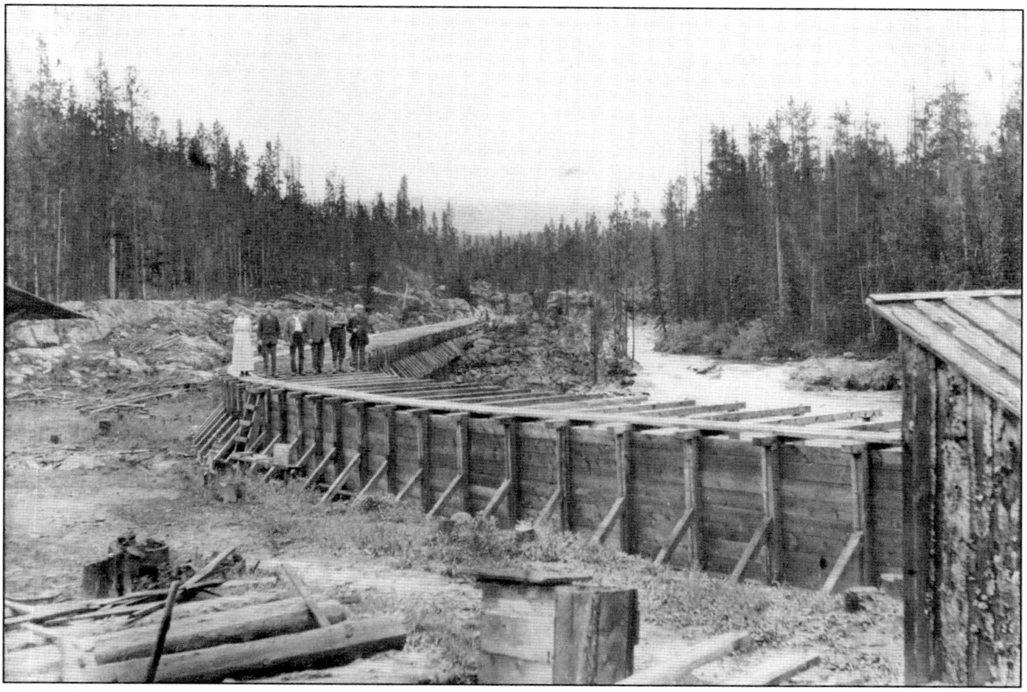

This photograph shows the upper section of the flume to the rock cut. The flume carried water from the Clark's Fork River to the turbines and rested on about 1,800 solid concrete and stone piers. It was a quarter of a mile long.

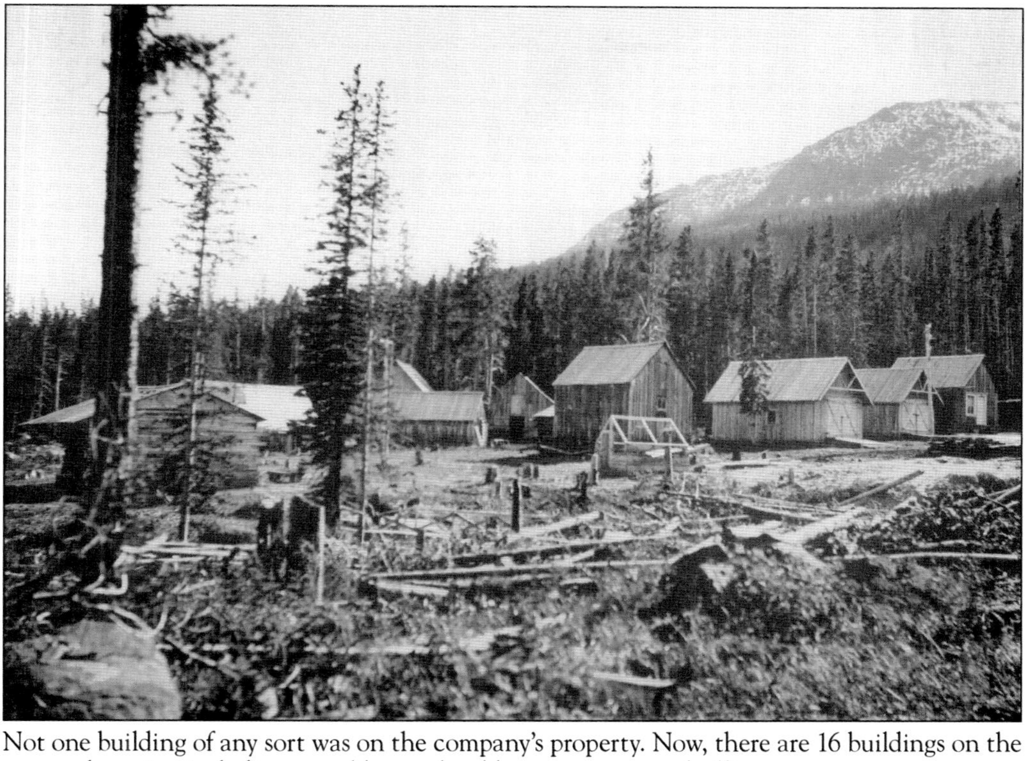

Not one building of any sort was on the company's property. Now, there are 16 buildings on the power plant site, including a cookhouse, bunkhouse, garage, and office.

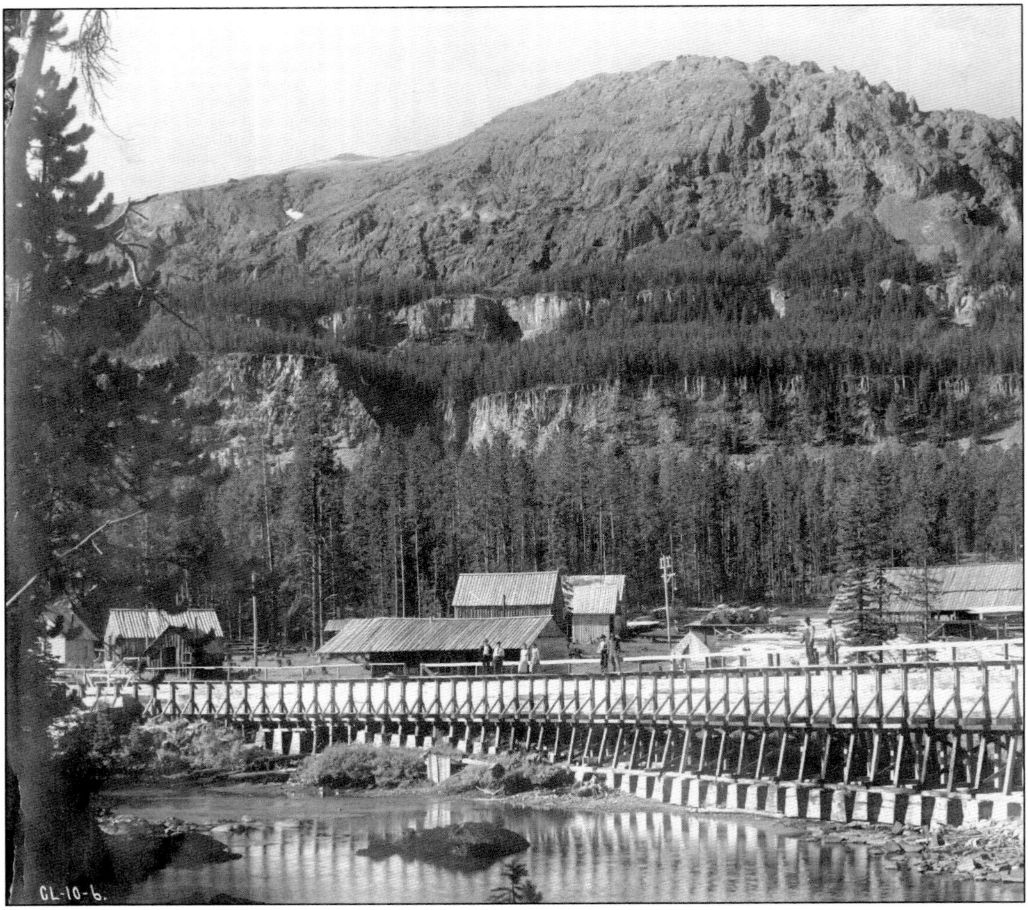

This photograph shows how extensive the project was near the power plant and lumber mills.

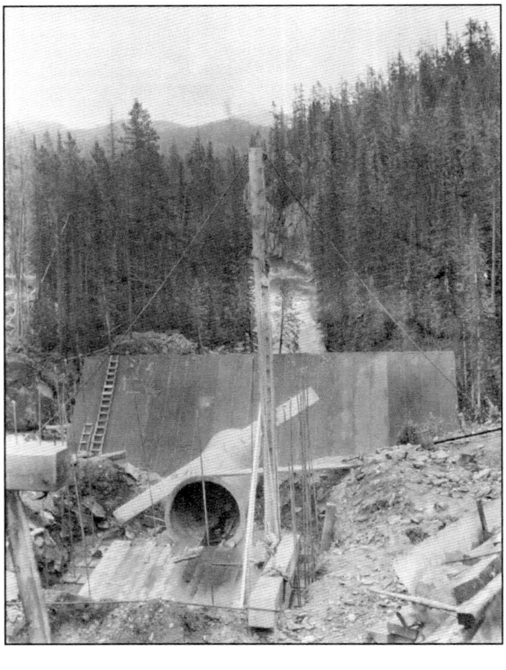

This image is looking down toward the roof of the powerhouse, showing the size of the penstock. The Clark's Fork River is behind the hydro plant.

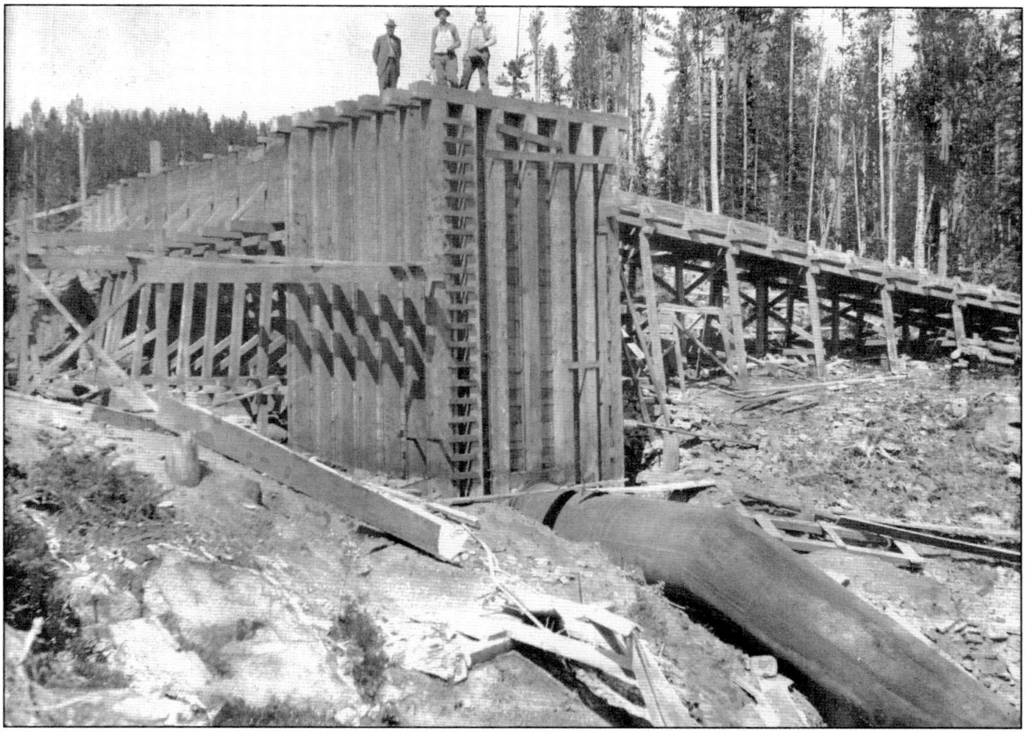

Western Smelting & Power Company manufactured more than 800,000 feet of timber used in the construction of the plant. During construction, these men are standing on top of the surge tank, where the water enters the turbines through a steel penstock, creating an effective head.

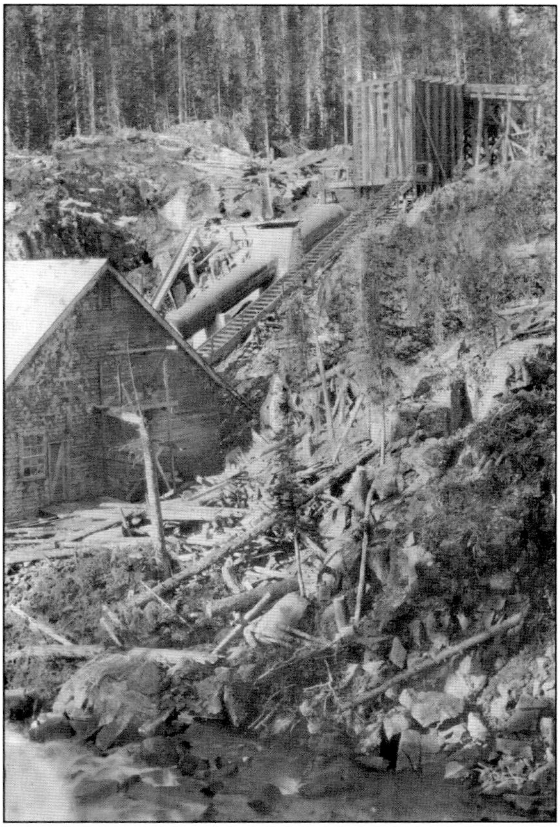

This image shows a view of the surge tank, penstock, and power plant from upstream.

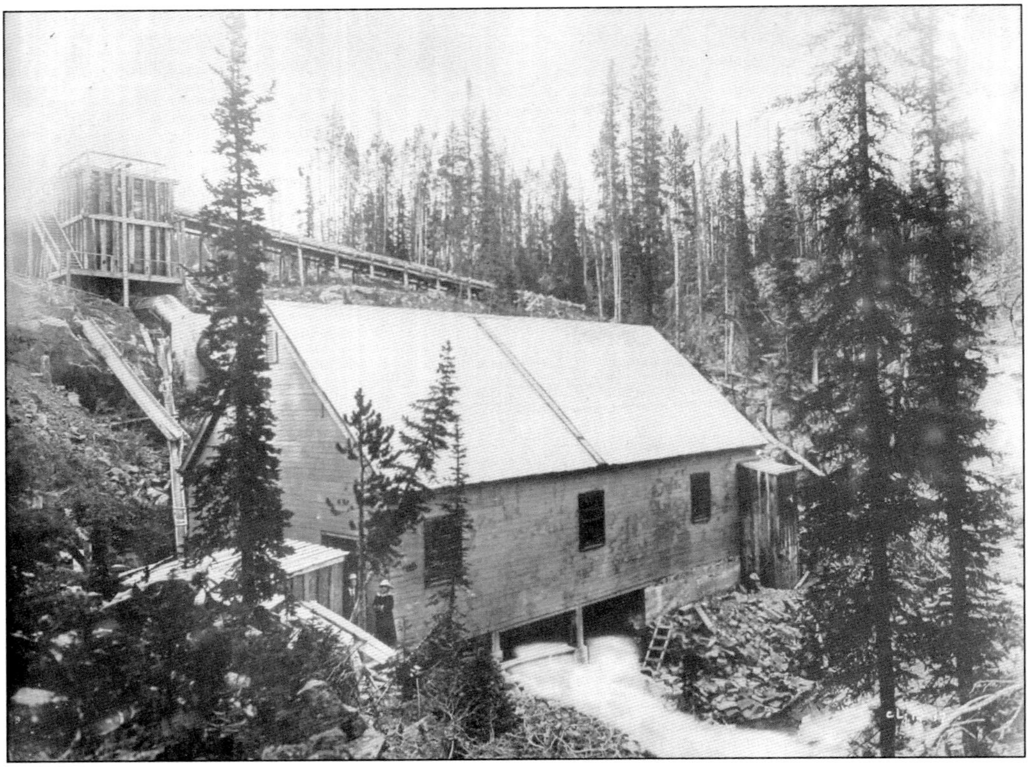

This is an exterior view of the hydroelectric plant after completion.

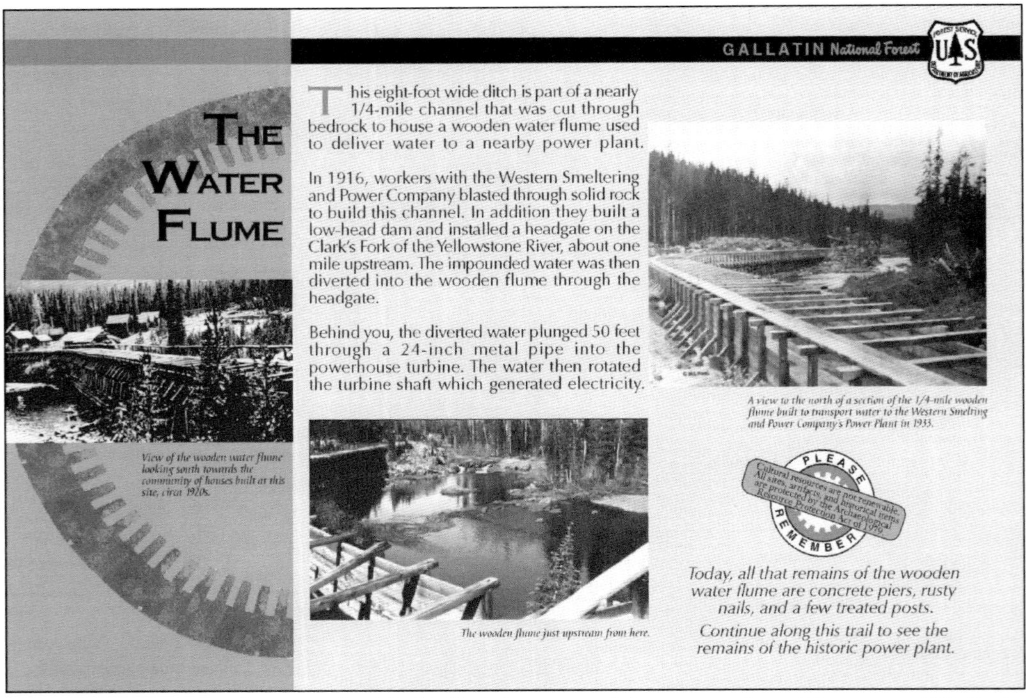

This interpretive panel is near the Clark's Fork location of the power plant. (Courtesy of Gallatin National Forest and Jane Ruckman.)

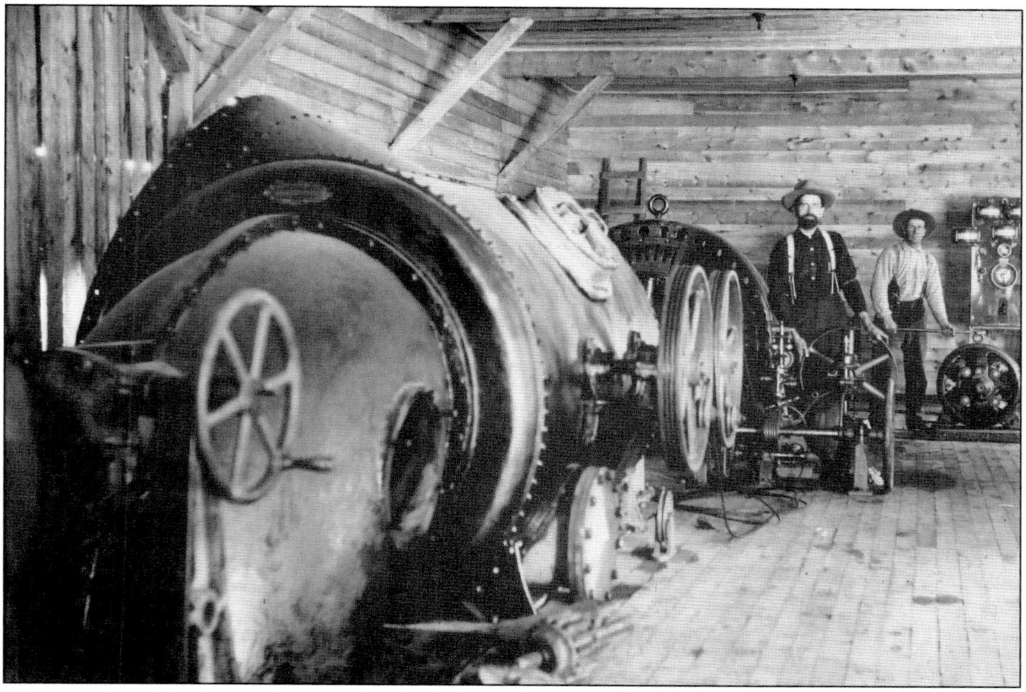

The interior of the hydroelectric plant had double horizontal turbines, wheels, generators, exciters, motors, transformers, and all other required electrical apparatus.

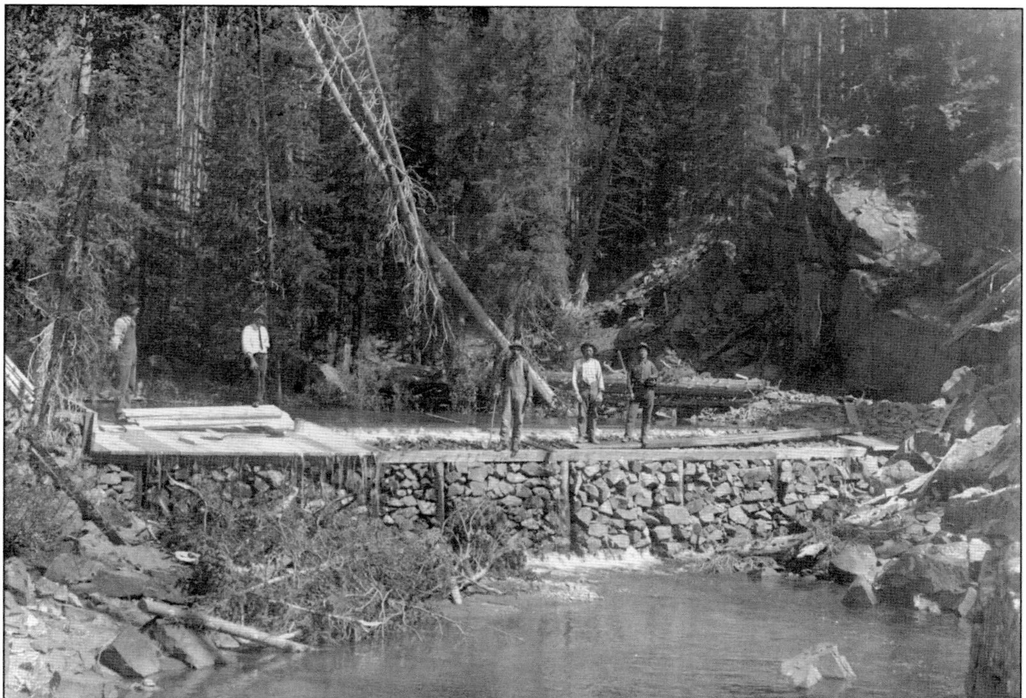

A low head dam impounded water, raised the level of the river, and diverted the water into the wooden flume. At times, the water was routed around this waterfall and used to turn turbines in the hydroelectric power plant before returning to its natural course downstream.

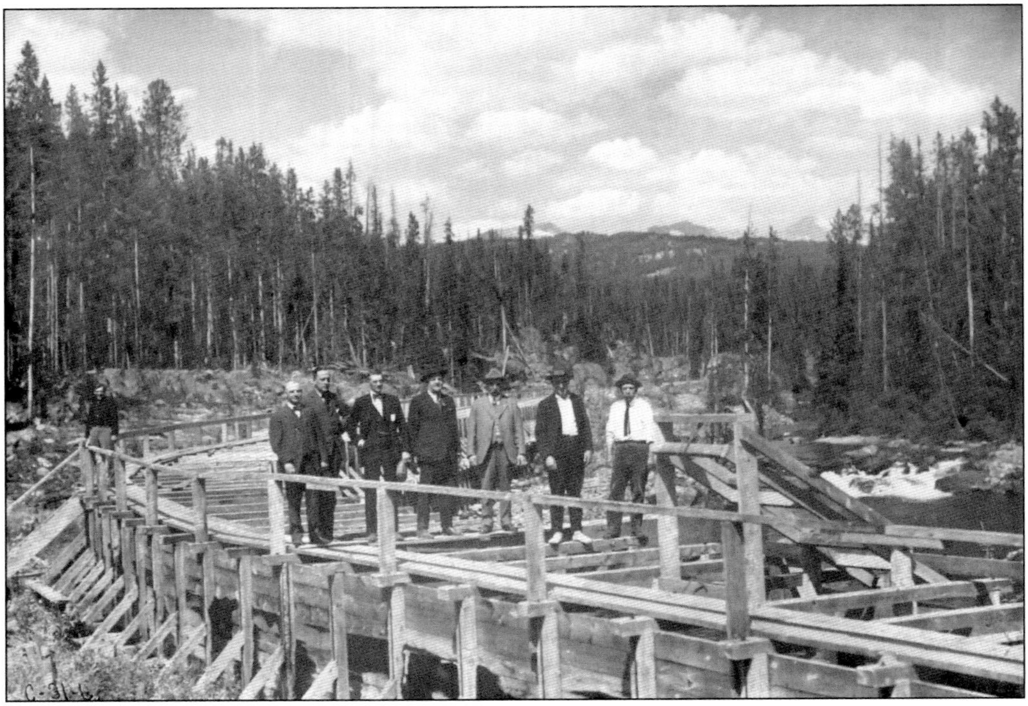

Written on the back of this photograph is " 'Power' and Flume W.S.&P. Co. [Western Smelting & Power Company] The men who make the wheels go around. Expert Fishermen!"

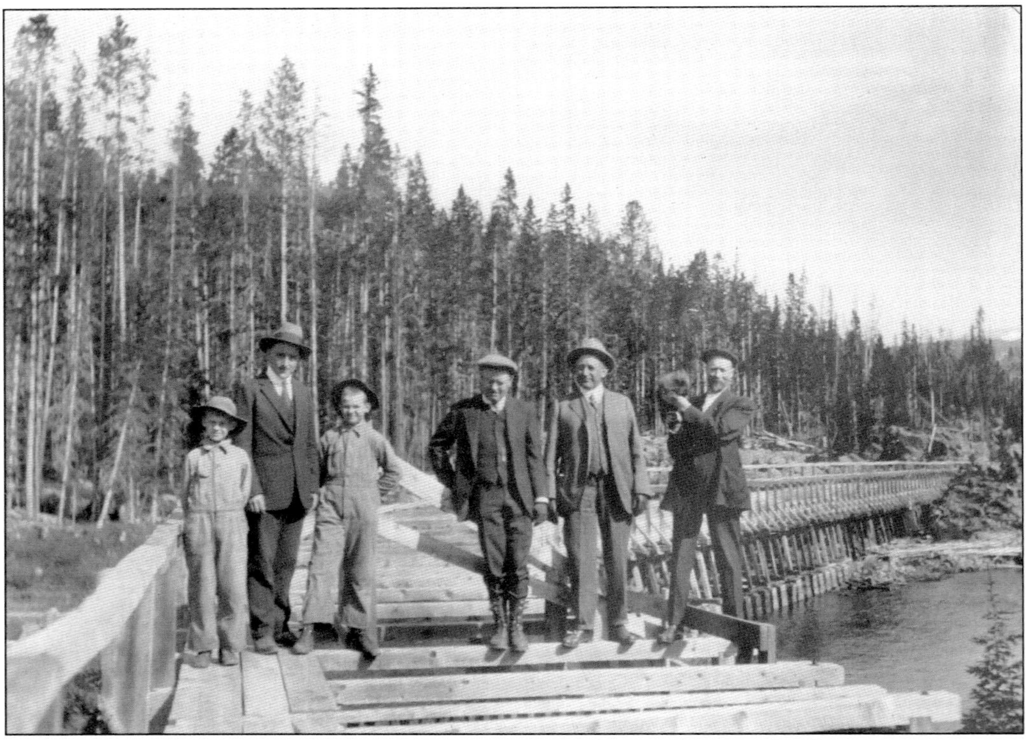

Bill Tanzer, the gentleman on the right, is standing with capitalists on the flume.

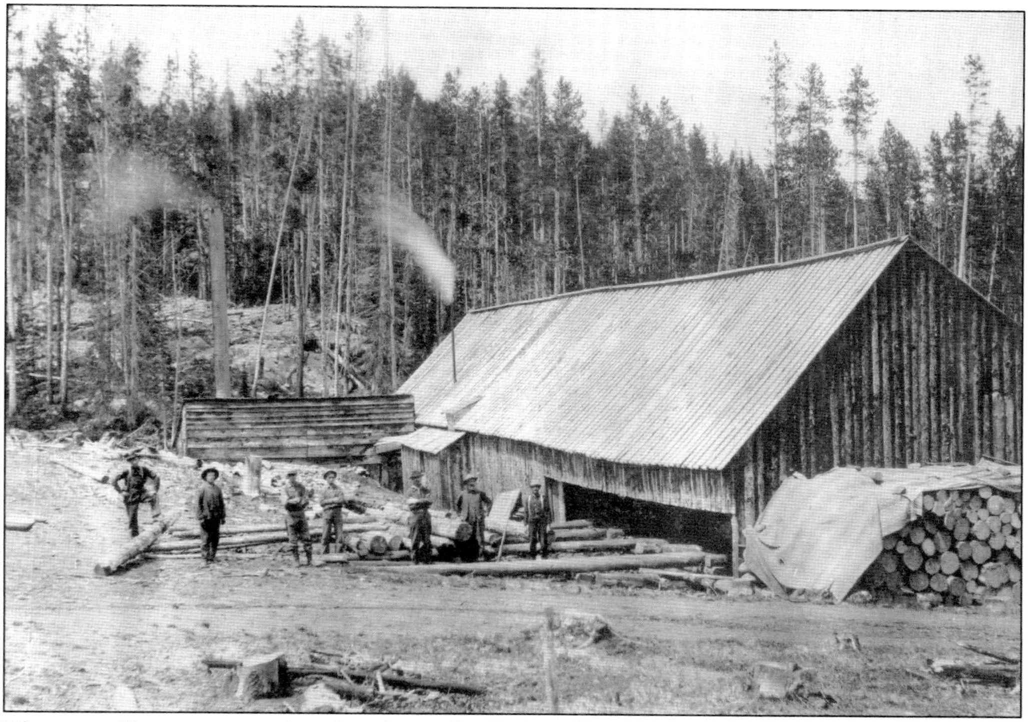

The sawmill was equipped with a four-side planer, edger, carriage with four head blocks, and circular saws. H.C. Reagan was the sawmill camp superintendent in this October 1916 photograph. Included in the picture are H.C. Reagan, Bill Tanzer, and Francis Brown.

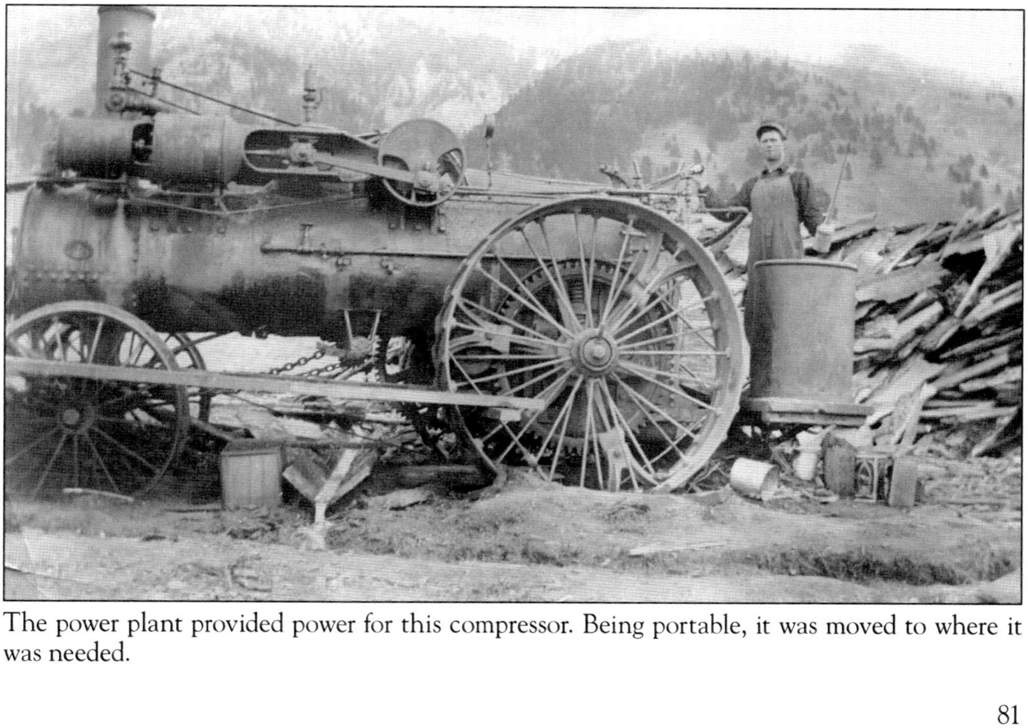

The power plant provided power for this compressor. Being portable, it was moved to where it was needed.

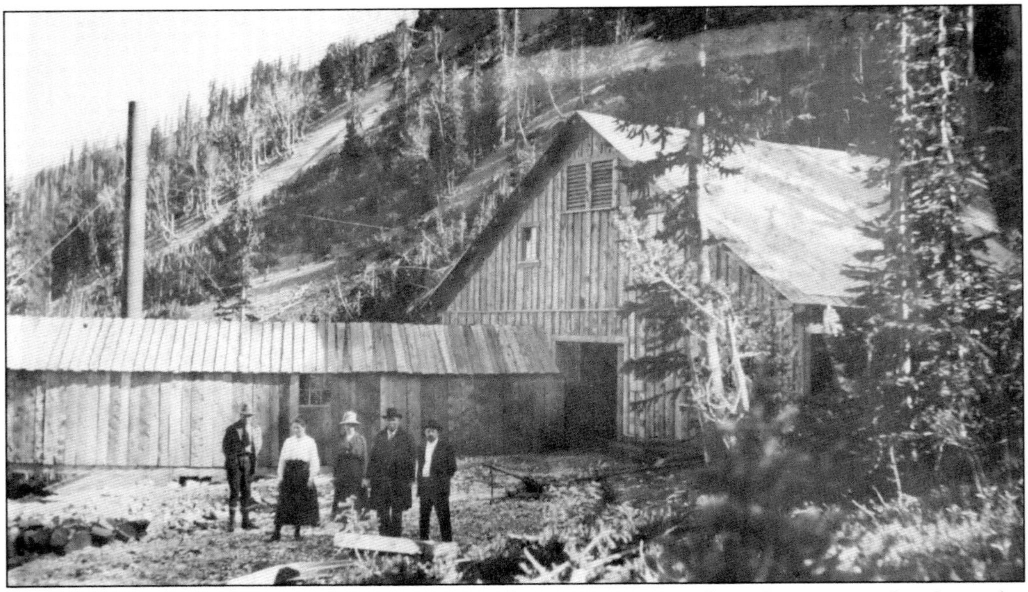

The Gold Dust plant was at the foot of Henderson Mountain, less than two miles from the smelter works. Expecting to tap the ore body exposed in the Homestake claim, nearly all of the company's development was building an 11,000-foot adit. By the end of 1925, little ore had been stockpiled or processed.

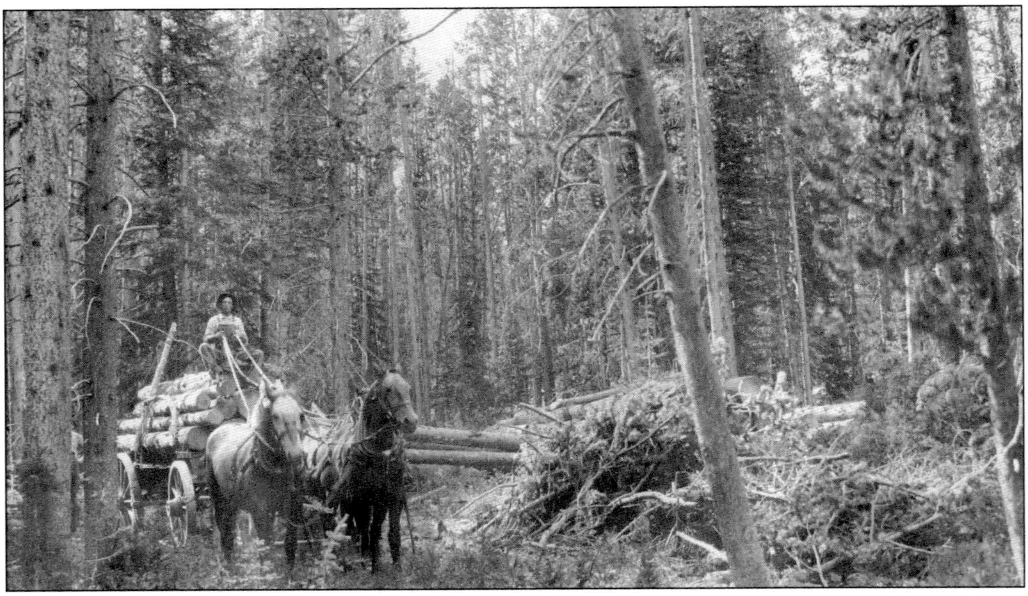

Pictured around 1915–1916, this worker is hauling logs to the sawmill.

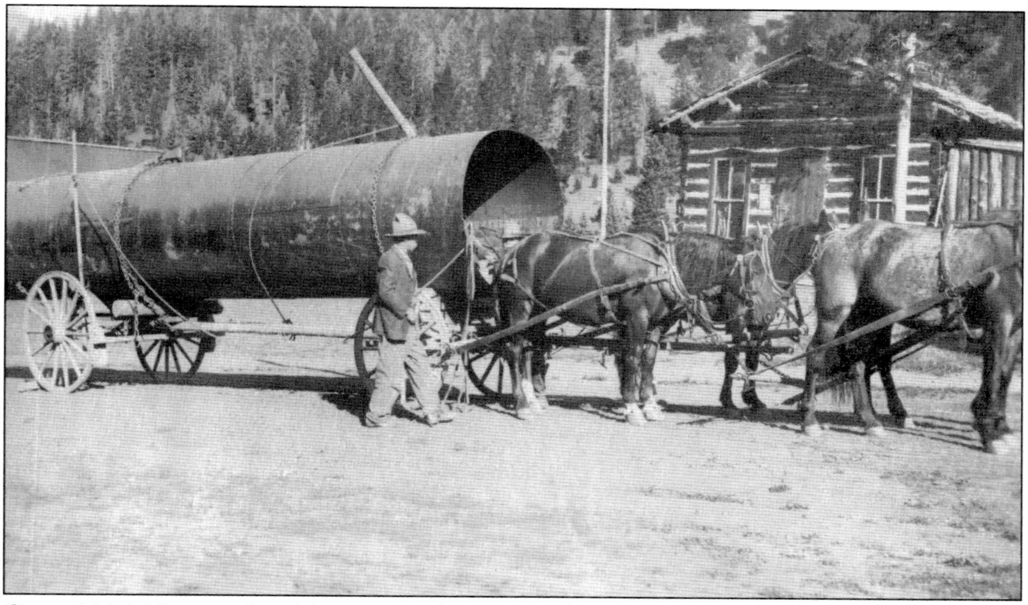

Over 600,000 pounds of heavy machinery had to be hauled over almost impossible wagon roads.

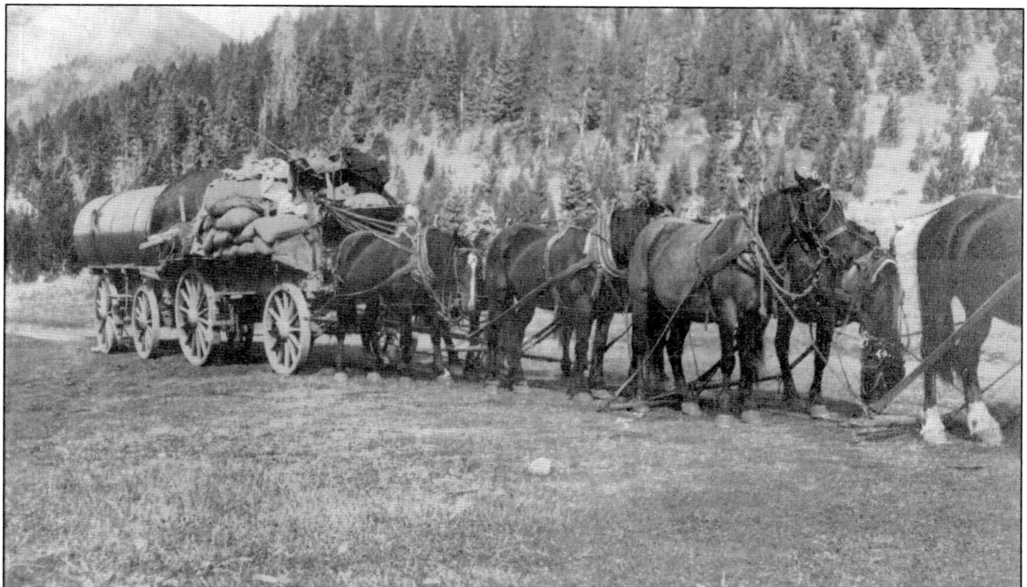

The smelter stack was freighted in from Gardiner around 1913–1914 by Bill Anderson's outfit. A single rope or leather line known as a jerk line was used to control these large teams hauling heavily loaded wagons.

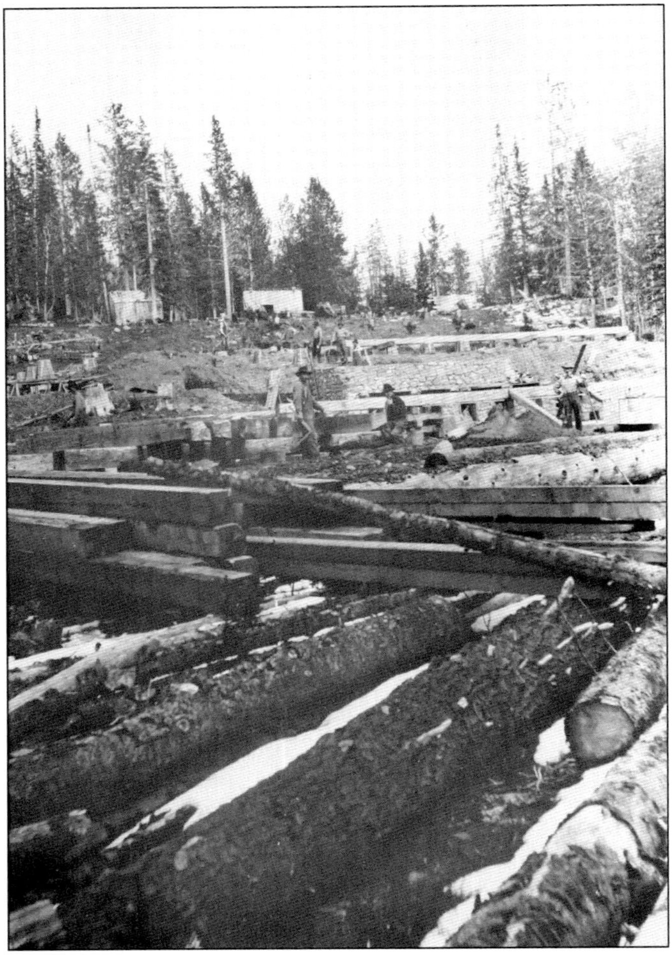

Trees have been cut and some stumps remain, but stone and wood foundations are being built.

These men are in management at the smelter.

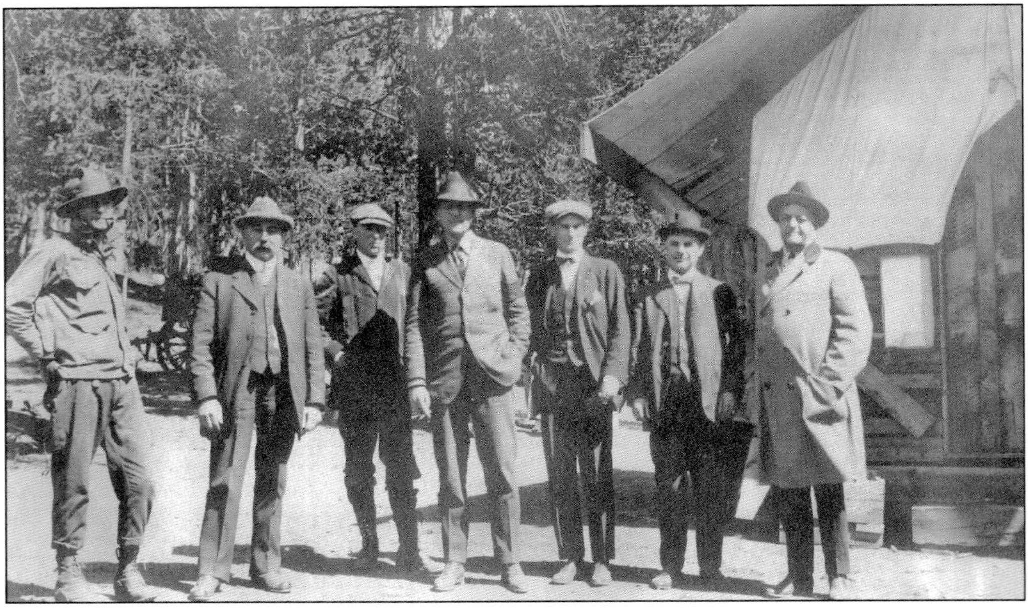

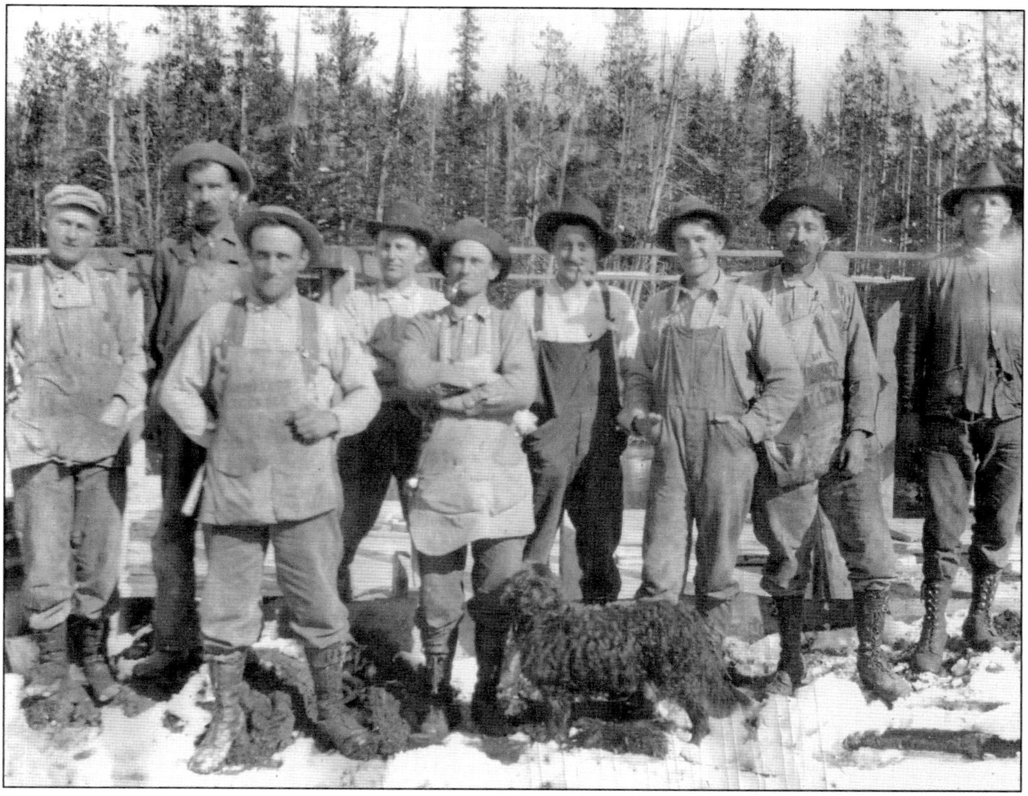

Employees at the smelter plant strike a pose in this undated photograph.

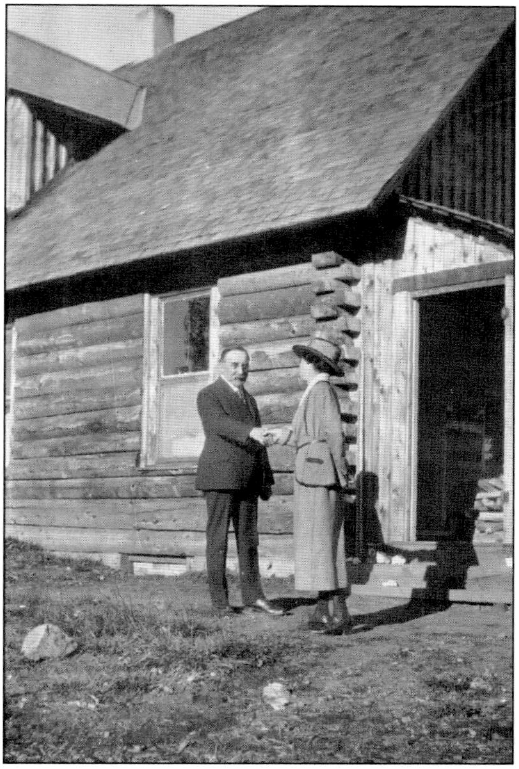

Dr. G.L. Tanzer appears in front of the superintendent's bungalow at the smelter camp.

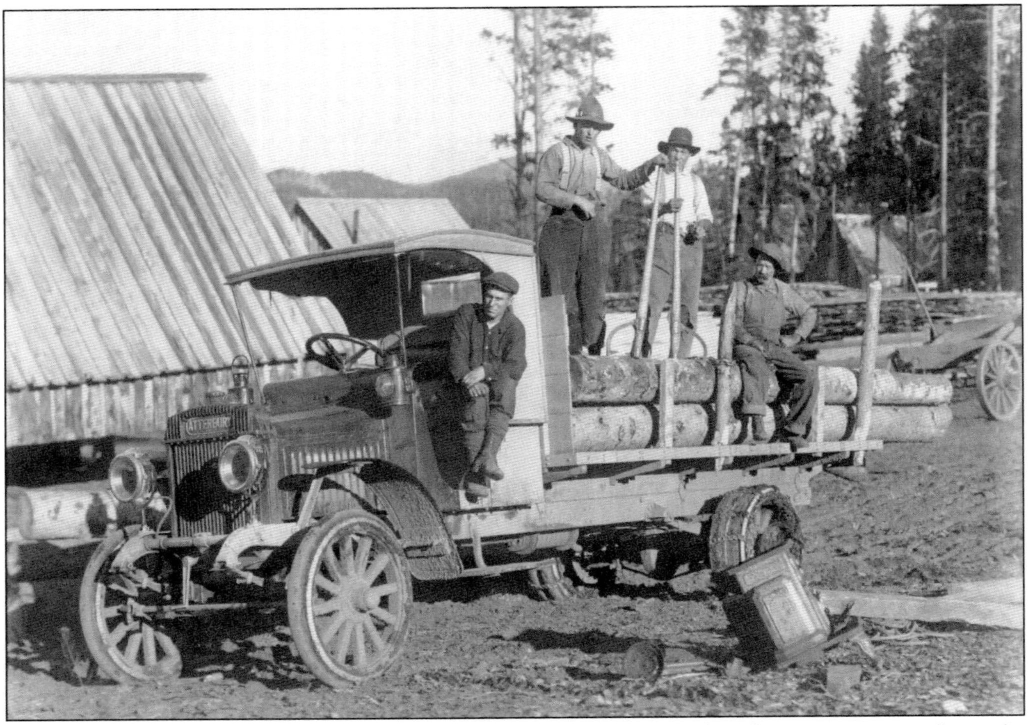

These men are hauling timber for Western Smelting & Power. Eby, the sawyer, is on the right and is missing his right hand.

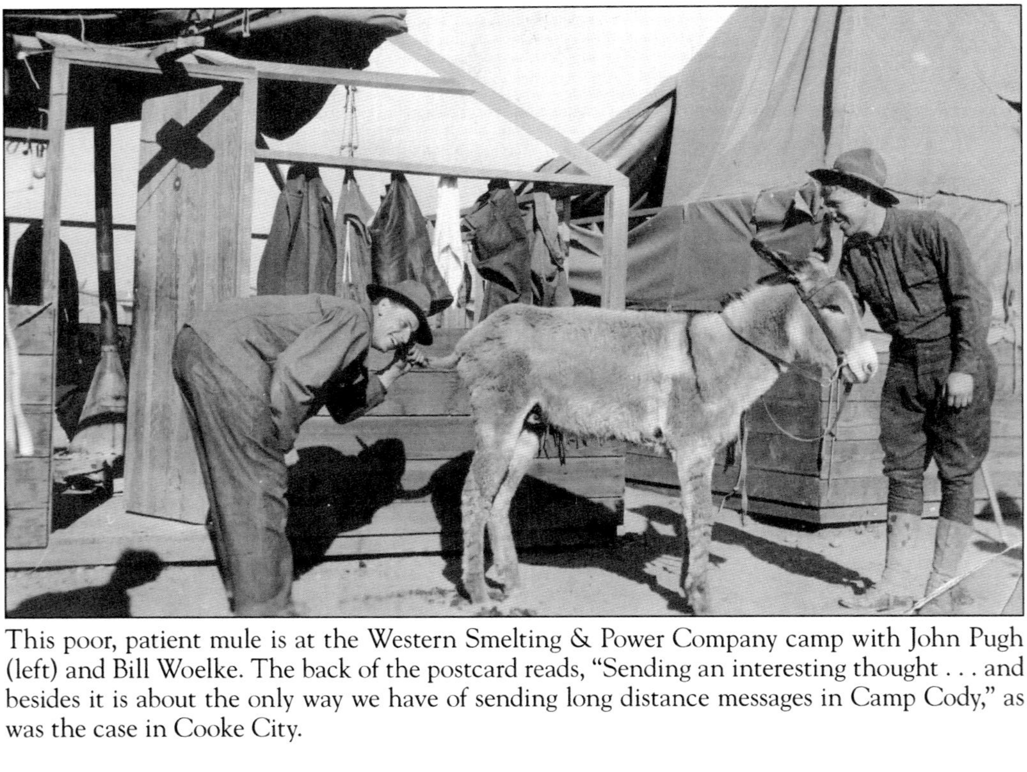

This poor, patient mule is at the Western Smelting & Power Company camp with John Pugh (left) and Bill Woelke. The back of the postcard reads, "Sending an interesting thought . . . and besides it is about the only way we have of sending long distance messages in Camp Cody," as was the case in Cooke City.

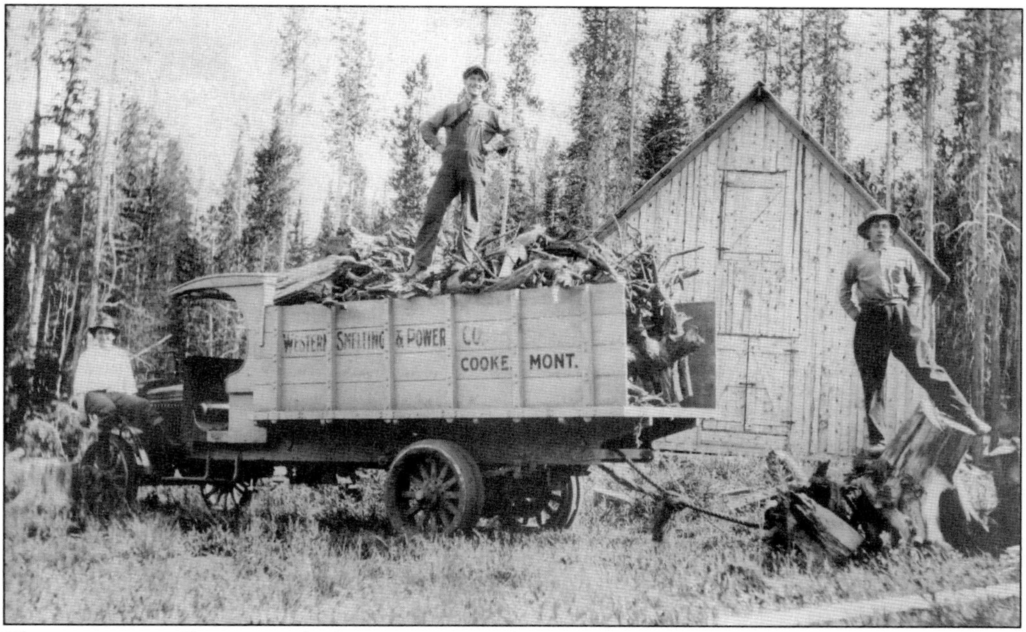

These men proudly pose with the Western Smelting & Power Company's auto truck.

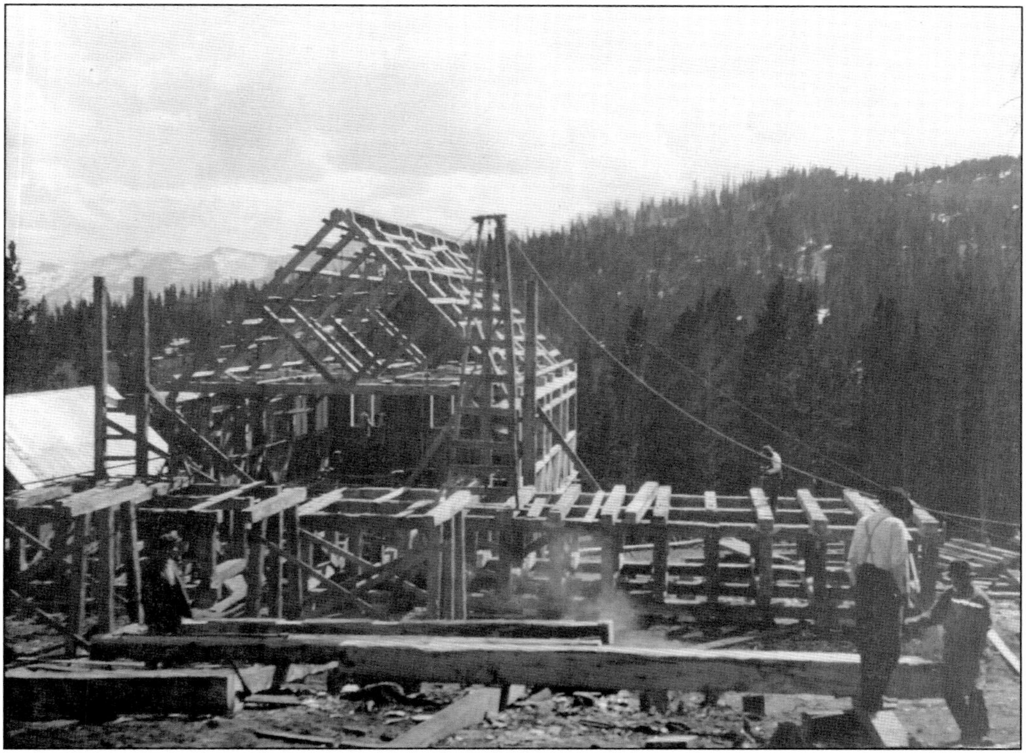
Shown here, eight ore bins with a capacity of 2,000 tons are being built. Today only a burned log gridwork foundation remains from this 30-foot-by-23-foot structure.

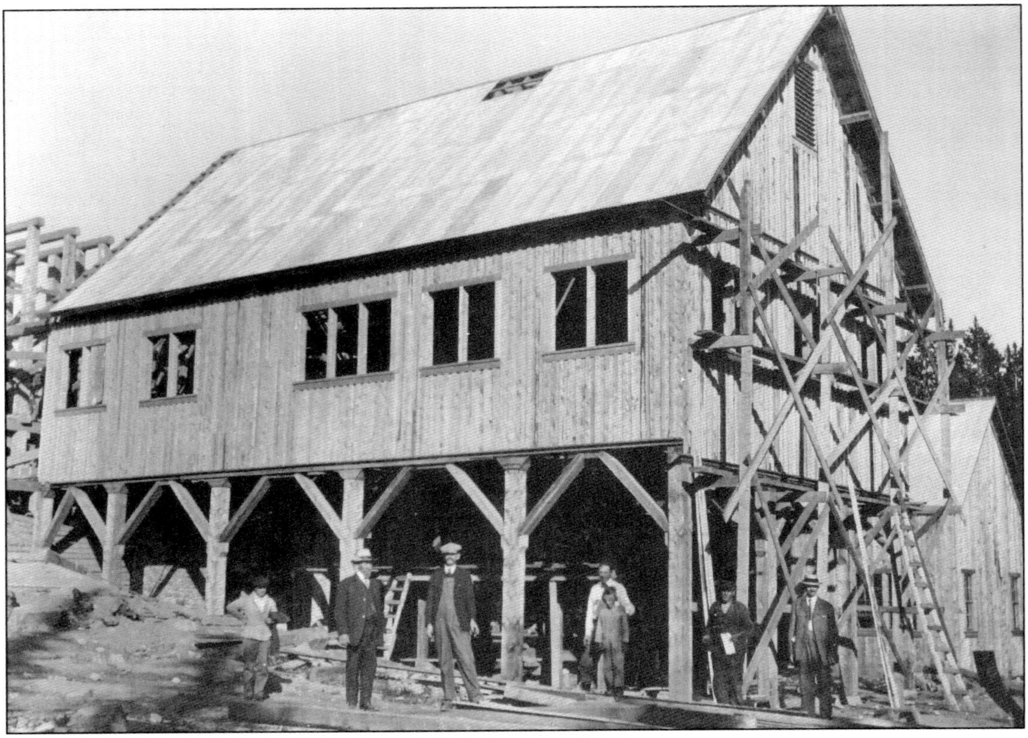

The blower house is pictured during its construction phase.

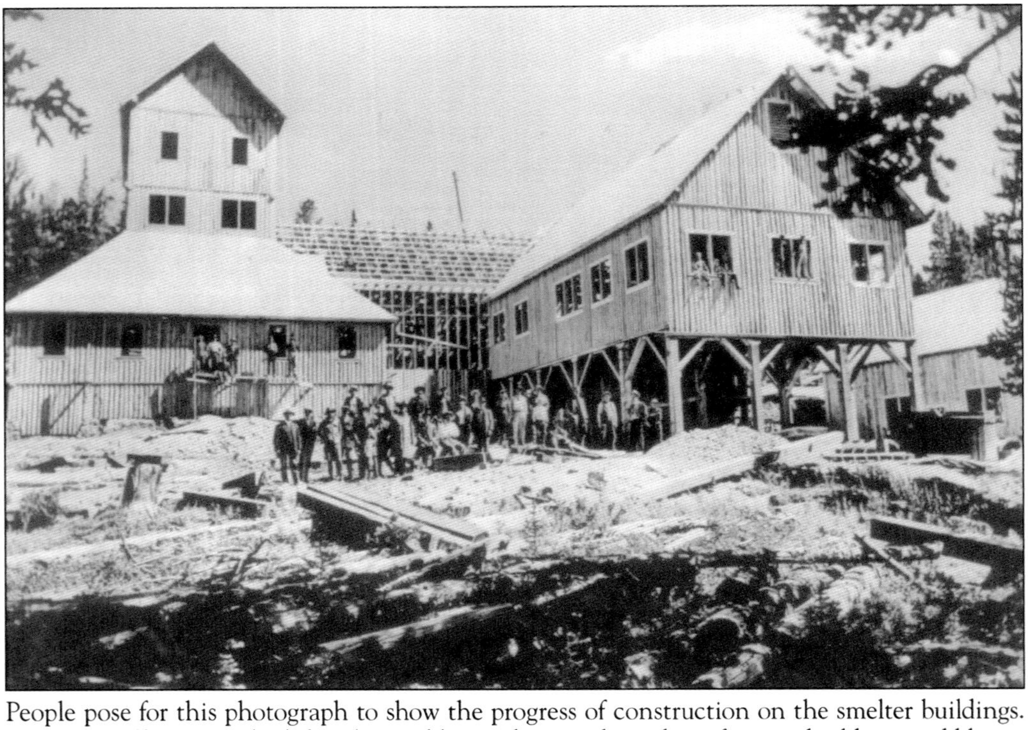

People pose for this photograph to show the progress of construction on the smelter buildings. The assay office is on the left. Also visible are the sampling plant, furnace building, and blower house. Individuals are standing and sitting on the window frames.

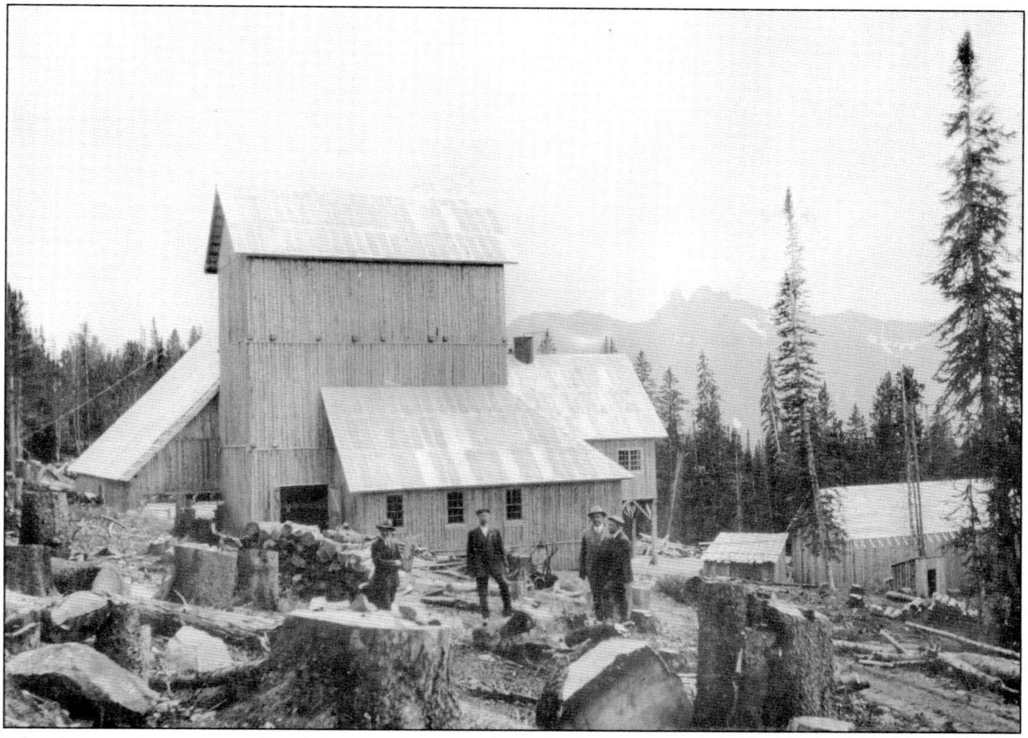
This image shows the side of the smelter buildings looking toward Index and Pilot.

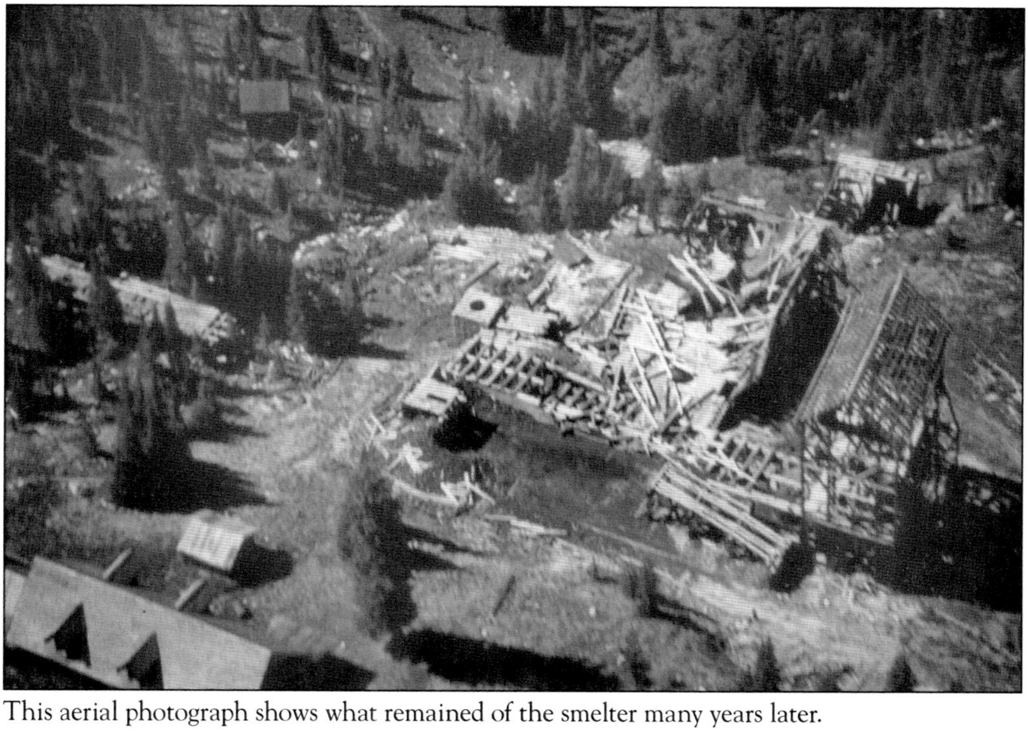
This aerial photograph shows what remained of the smelter many years later.

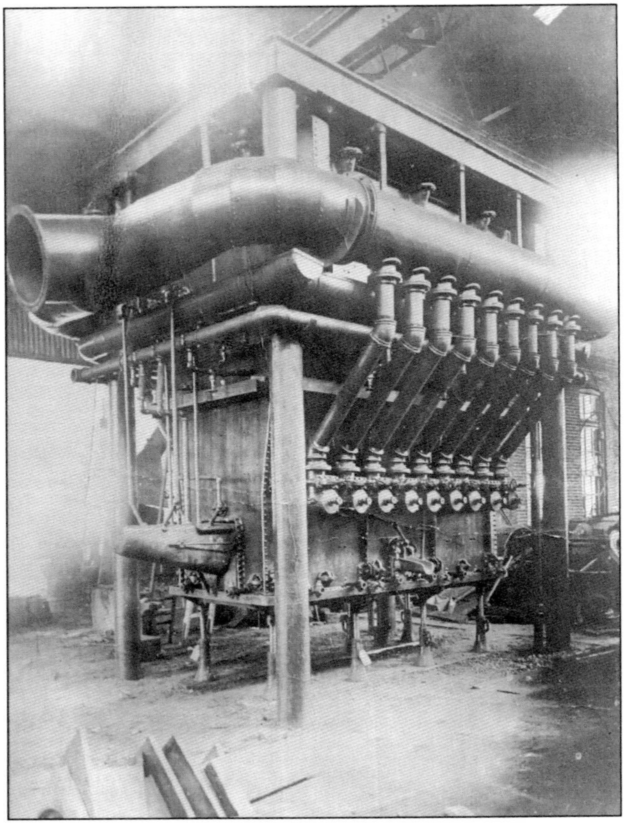

This building is equipped with a superstructure. The copper blast furnace is of modern design.

The steel hood of the blast furnace was connected to a 25-foot-long stack for downtake.

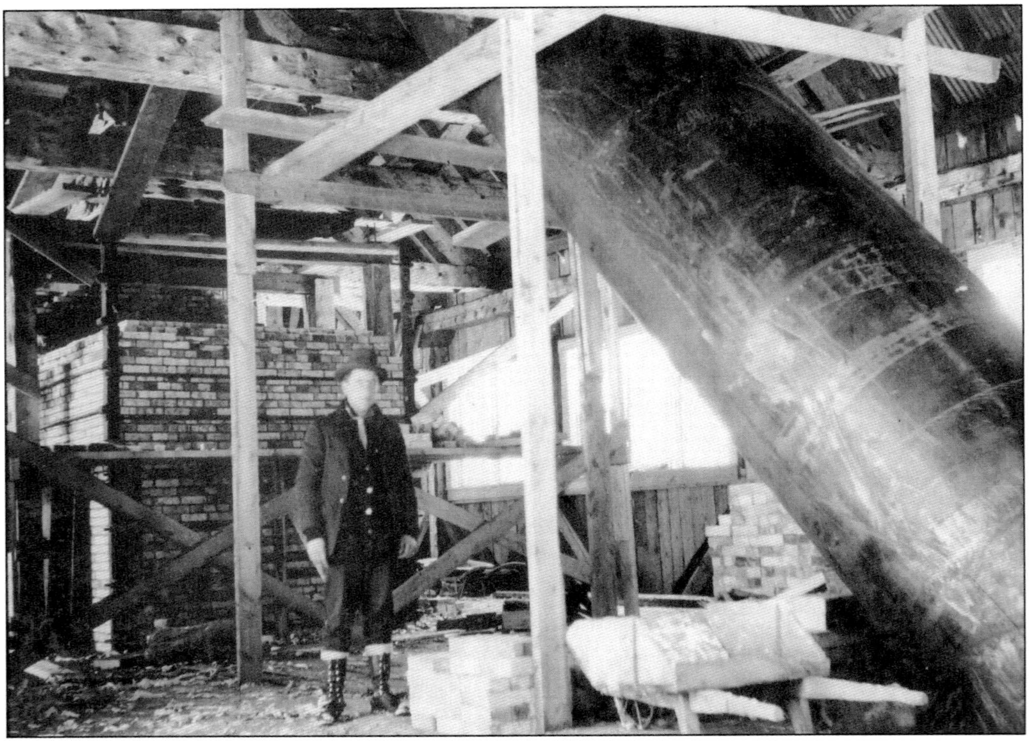

Bill Tanzer and smelter superintendent Francis Hancock appear on the charger floor along with smelter ore bins built of 12-by-12 timber.

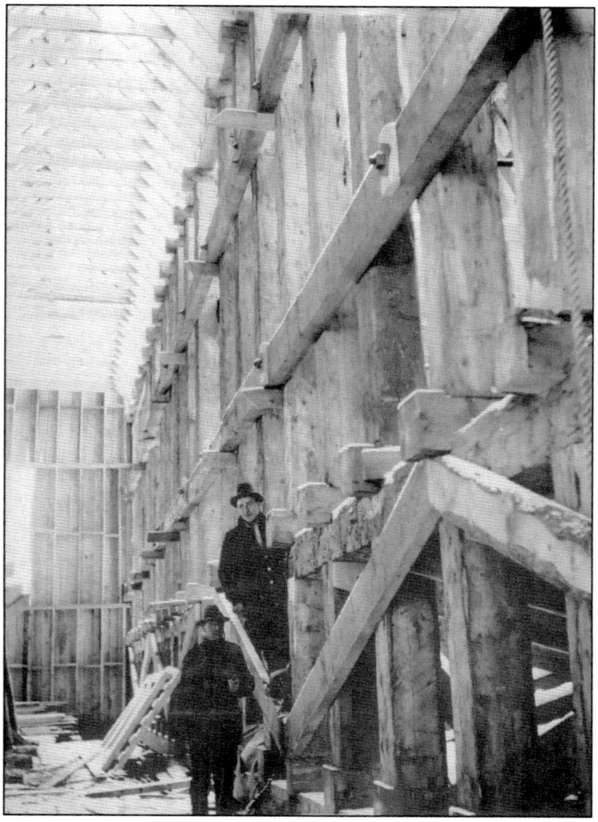

In 1916, the company received the concession from the Department of the Interior permitting the passage of auto trucks through the Yellowstone National Park from Cooke City to Gardiner, a distance of 60 miles. This auto truck and seven-passenger Packard touring car belonged to the Western Smelting & Power Company. Pictured from left to right are unidentified, Bill Woelke, Eby the sawyer, and Bill Tanzer.

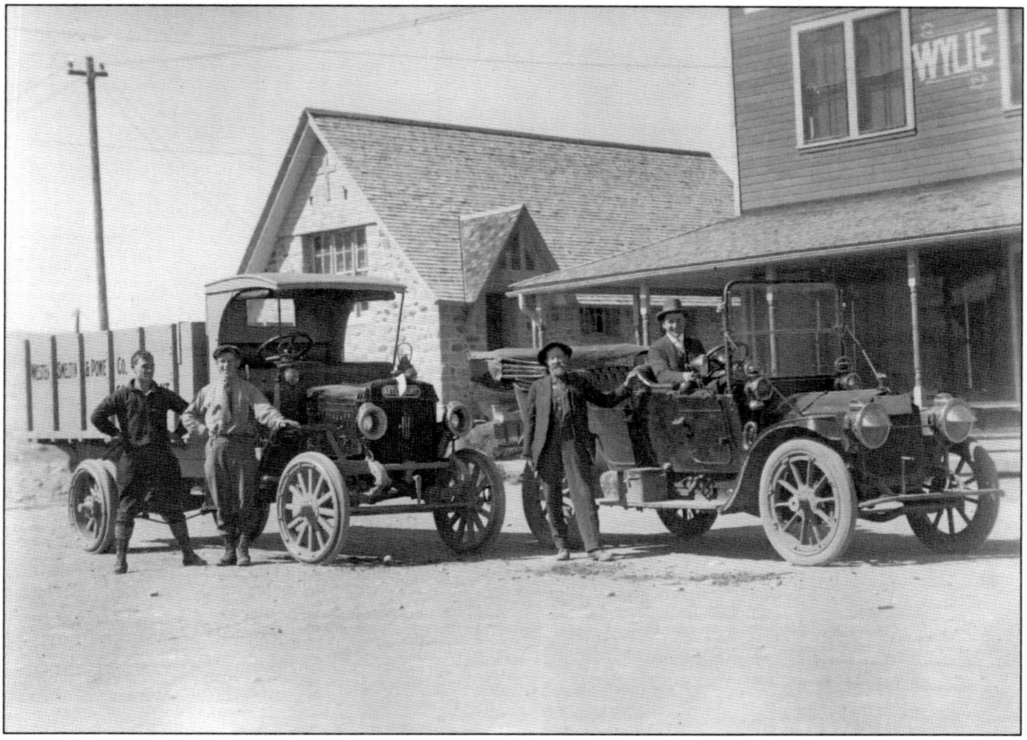

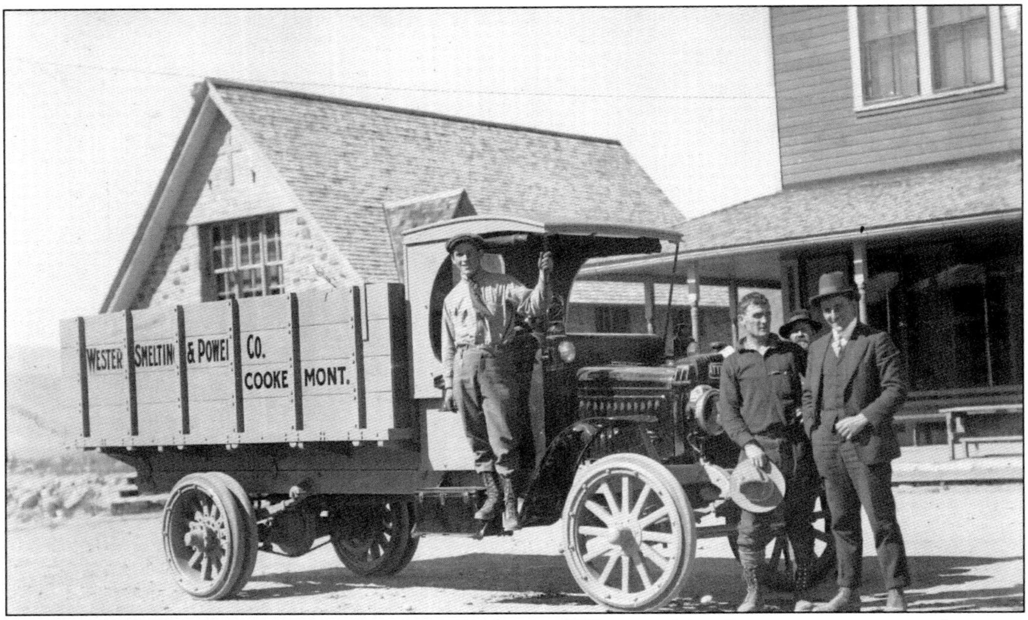

When the new smelter was in operation, trucks would be used for the transportation of copper matte until better means were available. The stone building with the cross on it is the Gardiner Community Church, built in the early 1900s. The Wylie Hotel, next door, is where guests were welcomed by the Wylie Camping Company before touring the park.

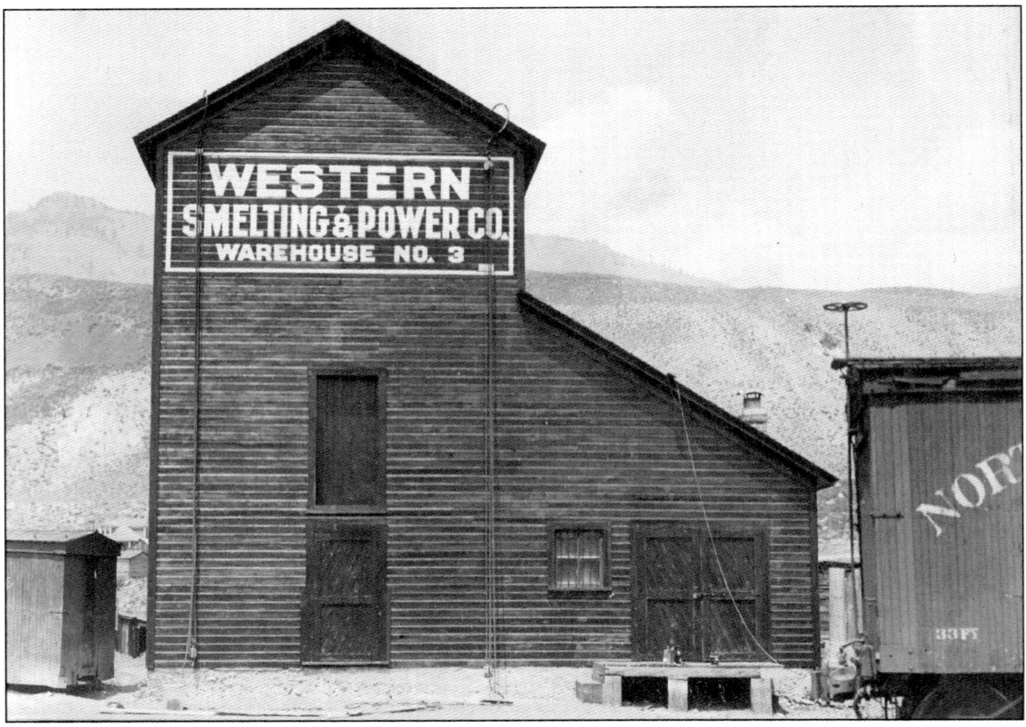

Western Smelting & Power Company's warehouse in Gardiner was opposite the freight depot.

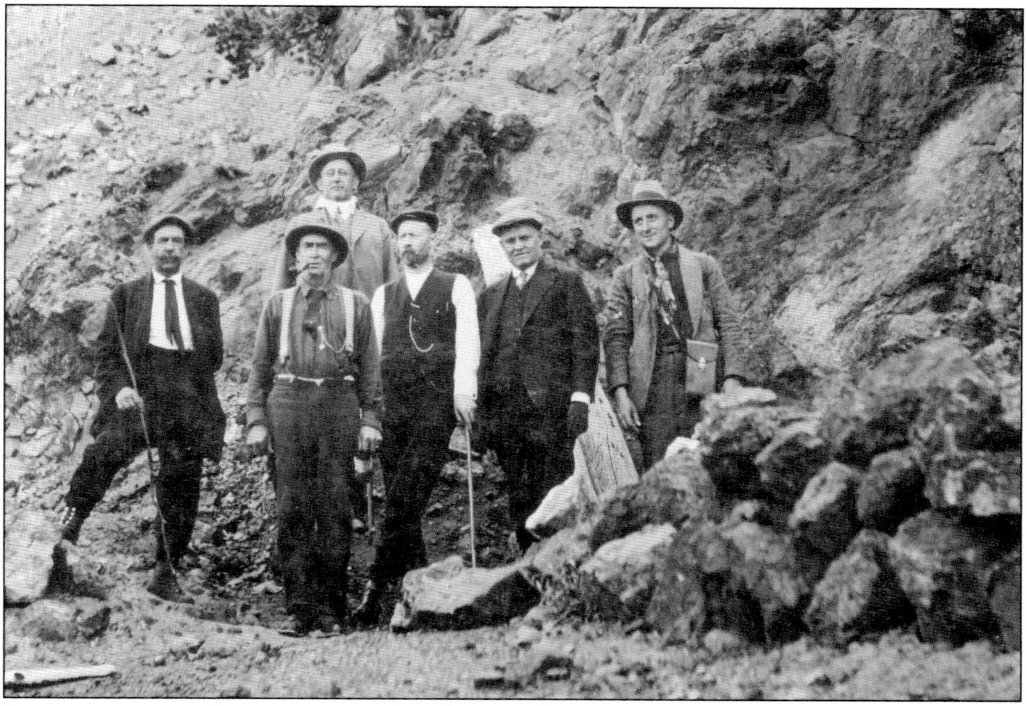

Immense outcrops of copper ore were found on Henderson Mountain, also known as the Mountain of Ore.

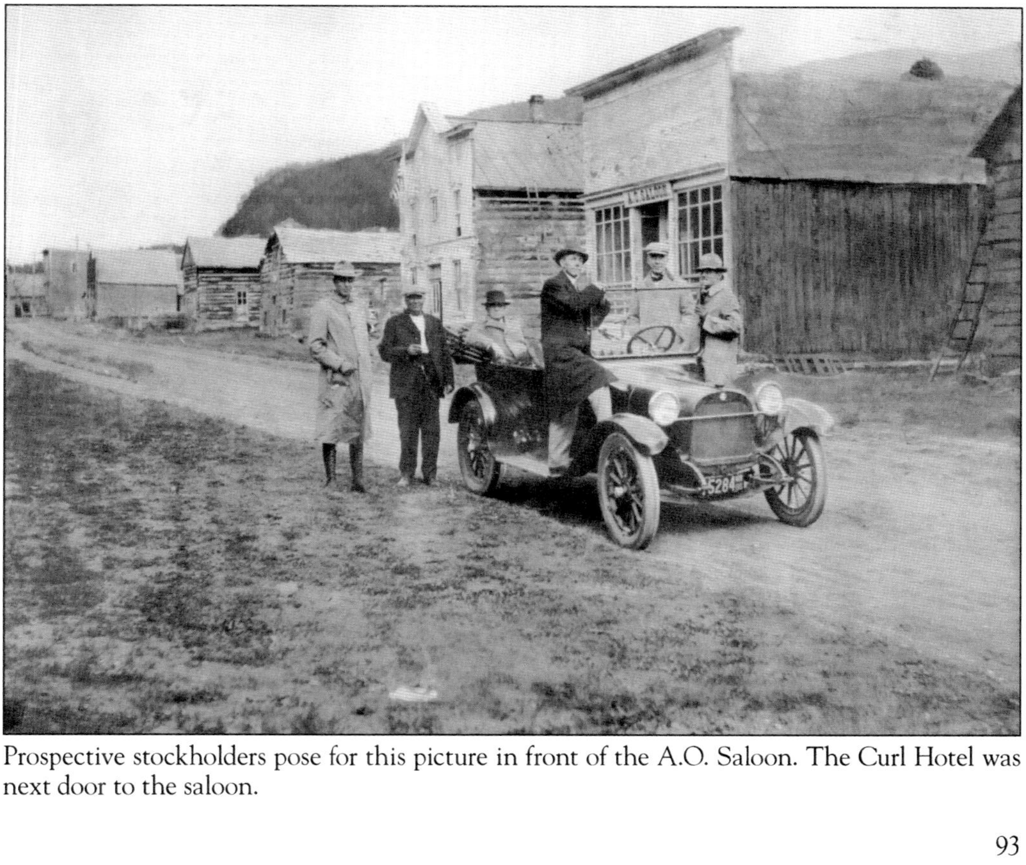

Prospective stockholders pose for this picture in front of the A.O. Saloon. The Curl Hotel was next door to the saloon.

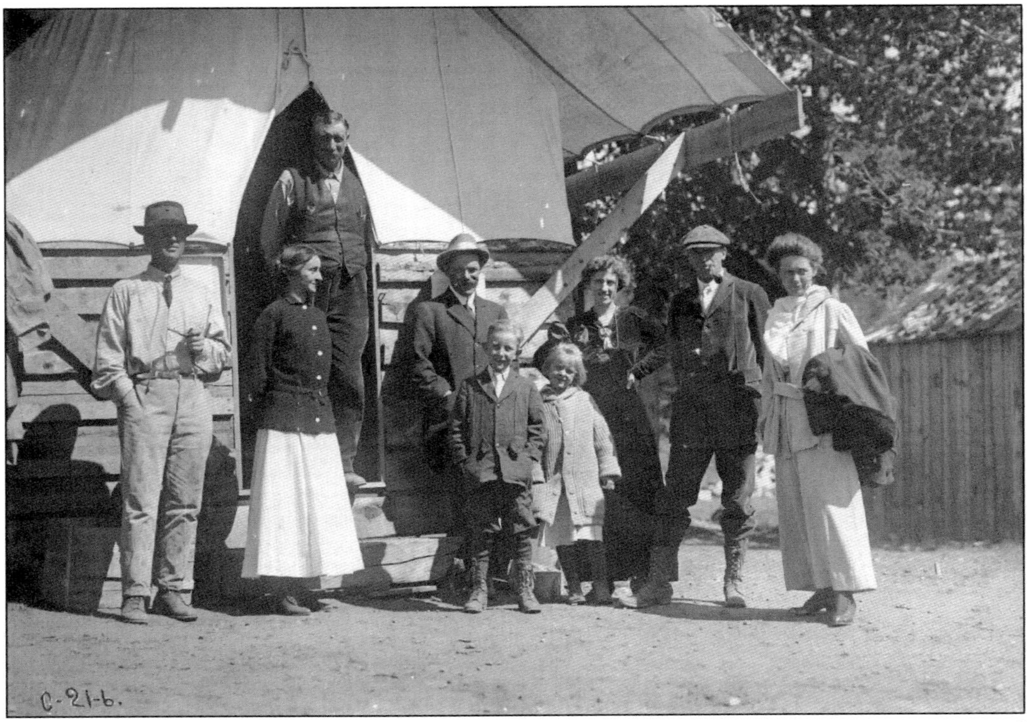

Prospective stockholders and families are entertained here. The Montana Company announced in the summer of 1907 that it would conduct an excursion to the wonders of the park in special carriages.

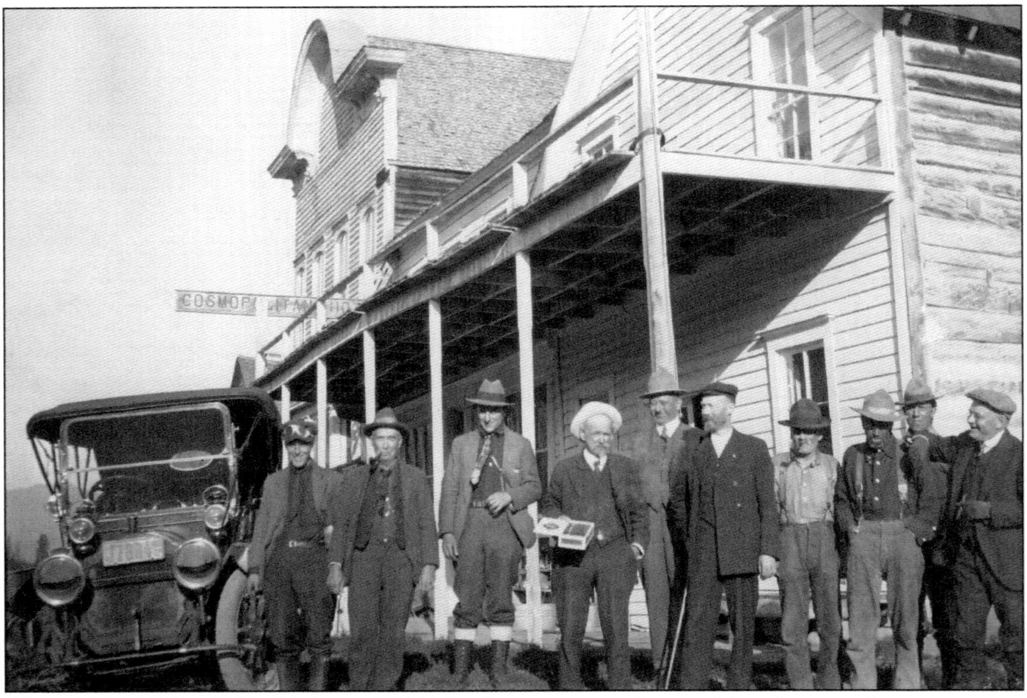

These men appear to be enjoying each other's company in front of the Cosmopolitan.

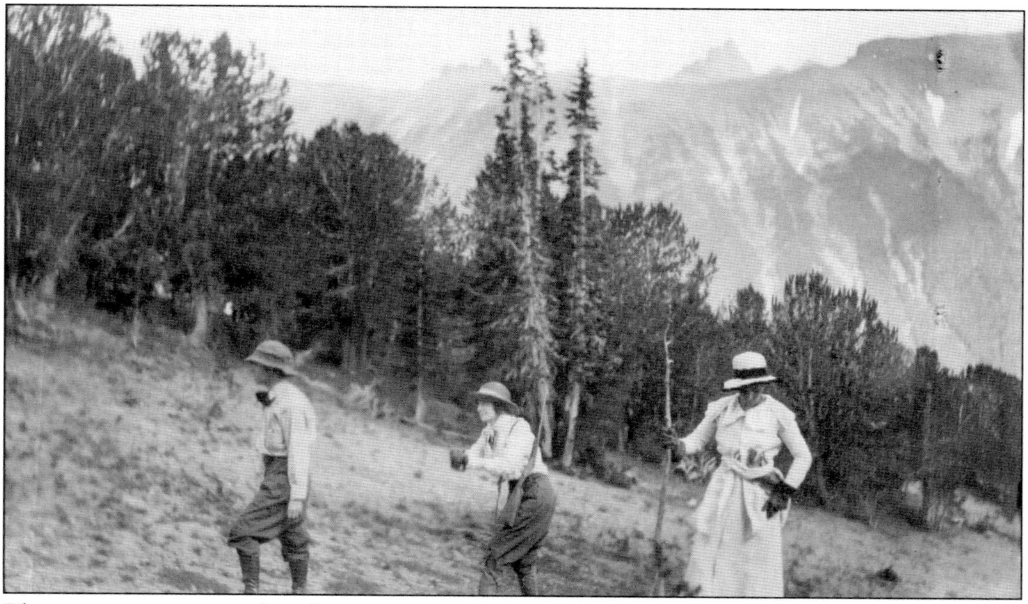

These women are armed in this primitive area: one has a rifle and one a pistol for protection.

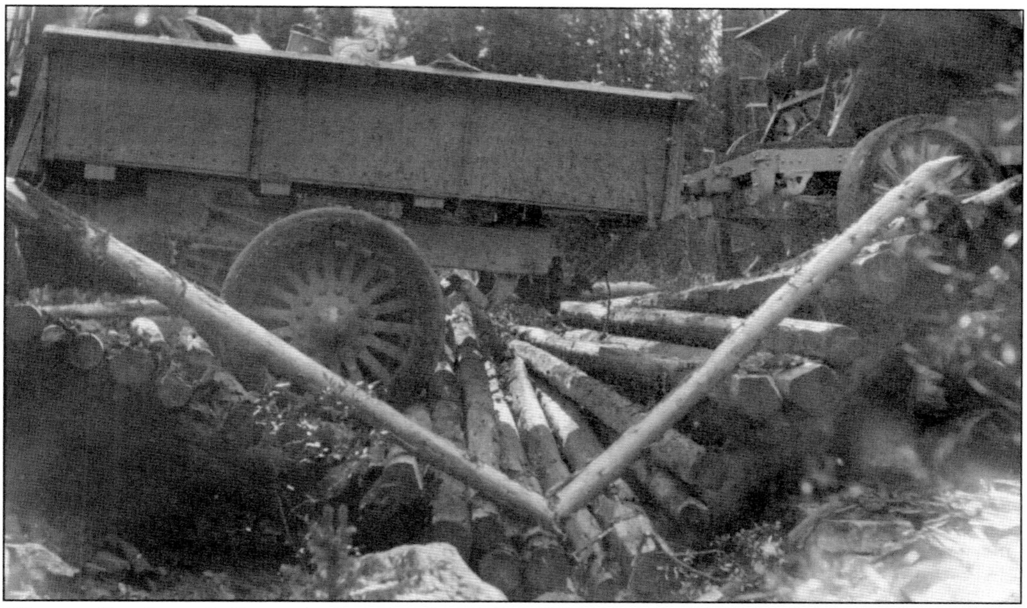

Western Smelting & Power built 14 bridges. Not all bridges were built to the demands of heavy trailers filled with ore.

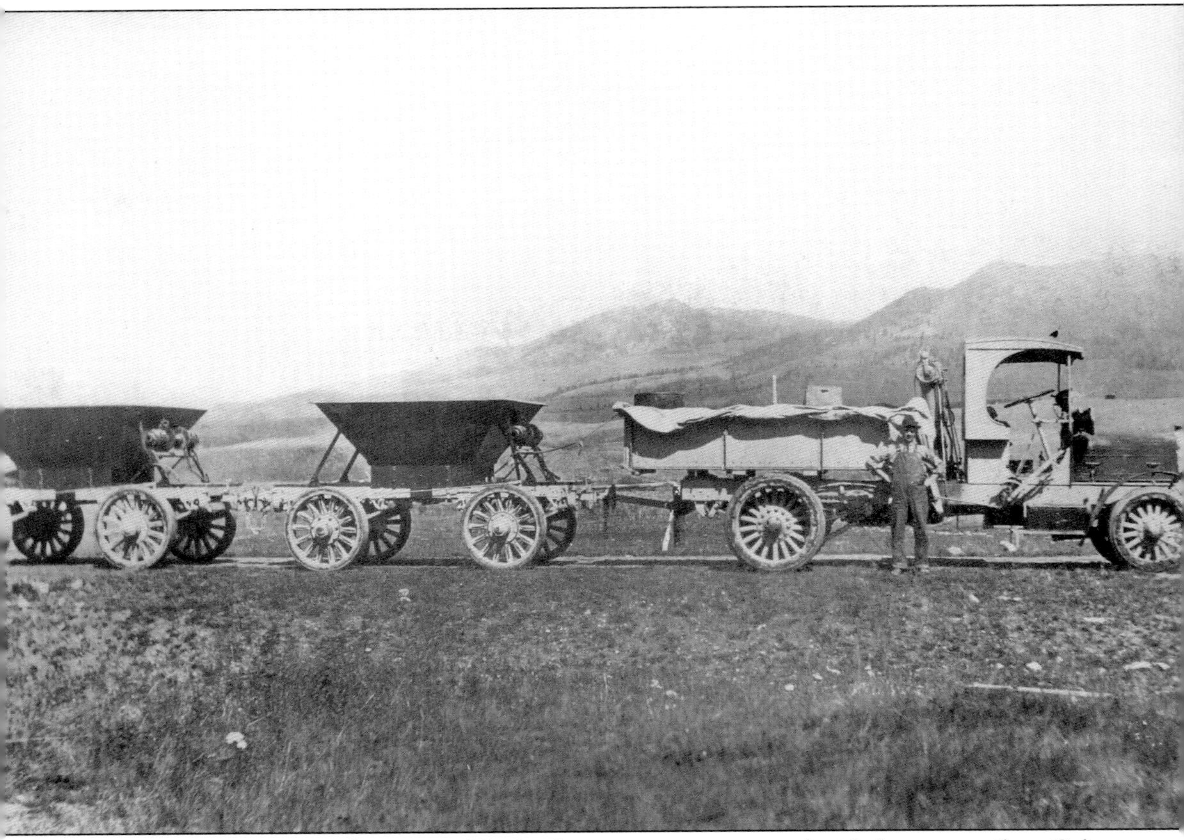

A fleet of five-ton trucks was used for hauling ore, some equipped with five-ton trailers. Others were used for transportation, supplies, and hauling fuels.

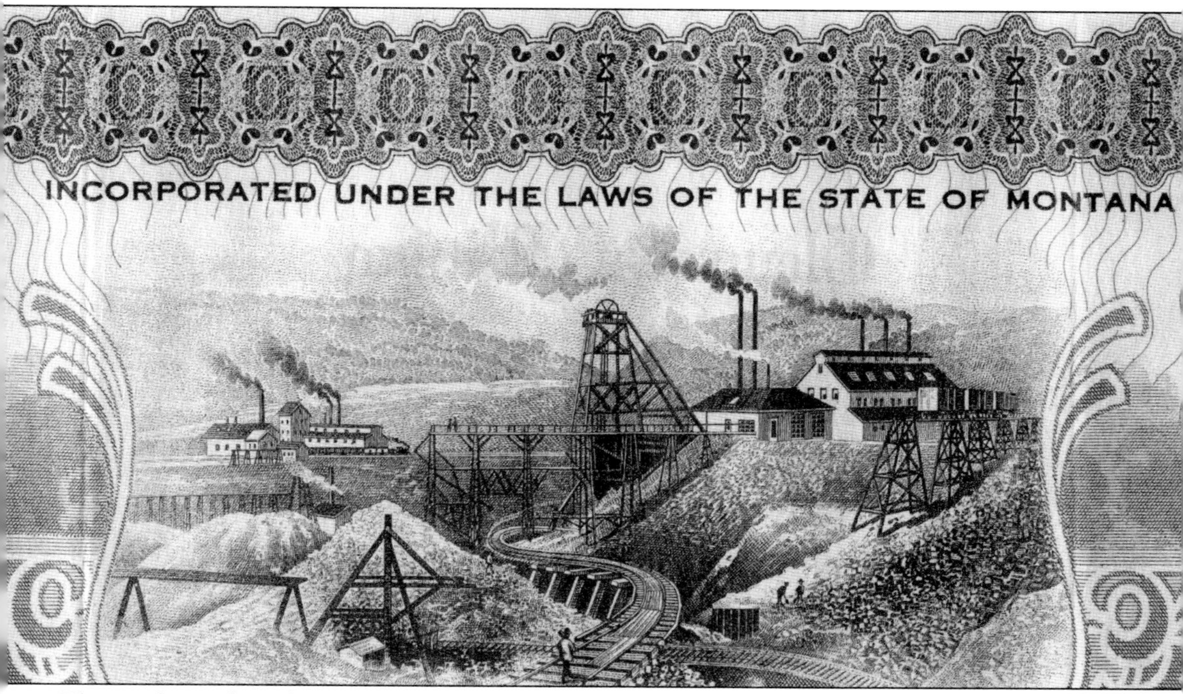

This stock certificate has a picture of smoke billowing out of stacks, evidence that the plants were operating. Dr Tanzer had said the smoke would be rolling from the stacks of the smelter soon. A fire was started in the smelter and weeds were burned so that the photograph showed the smoke rising from the stacks, which was evidence that the plants were in full operation. However, the smelter was never fully operational. Equipment was found still in crates that had never been installed. Dr. Tanzer went back east and then disappeared. There were few outstanding bills. The stockholders lost their investments, or were they speculations?

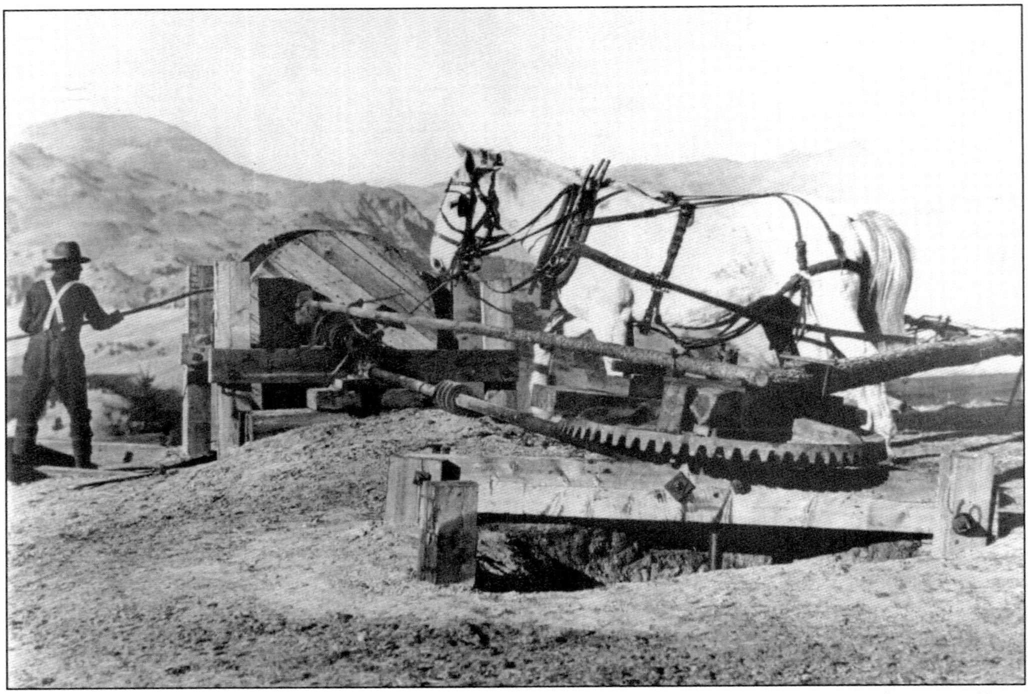

After the mine in Aldridge closed, the big tram cables were dismantled. Joe Narlow is rolling up the Foster Line cable with a turntable he built called a team hoist. The cable was bought by G.L. Tanzer of Cooke City for the Western Smelting & Power Company's tramway. Plans were made to use tramcars for automatic loading and dumping of ore. (Courtesy of Yellowstone Gateway Museum.)

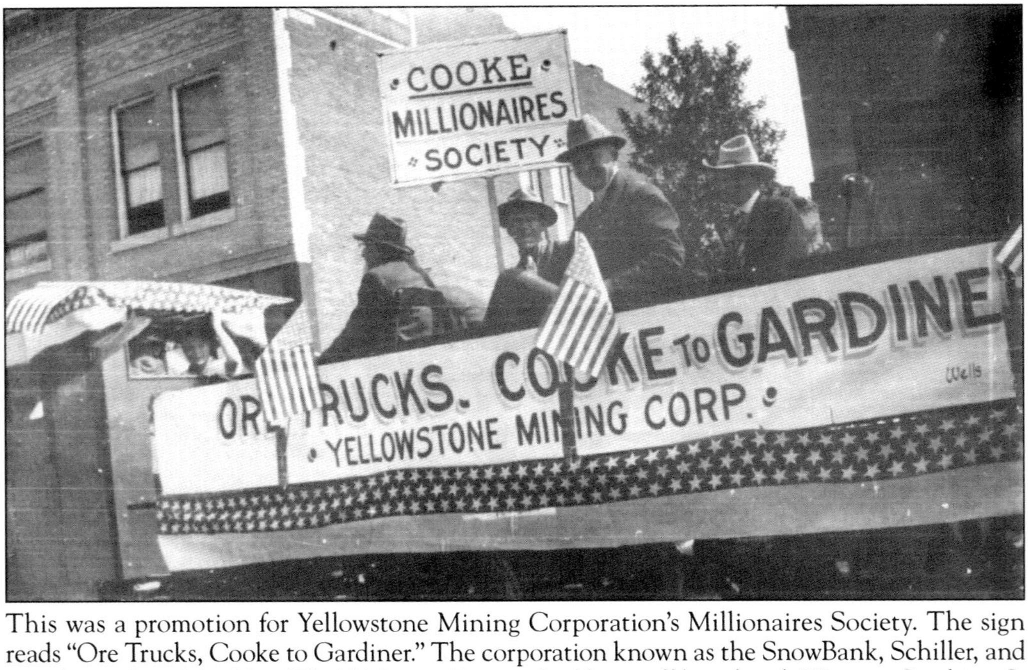

This was a promotion for Yellowstone Mining Corporation's Millionaires Society. The sign reads "Ore Trucks, Cooke to Gardiner." The corporation known as the SnowBank, Schiller, and Goethe Lodes of the New World Mining District had been affiliated with Western Smelting & Power Company.

Seven

Goose Lake

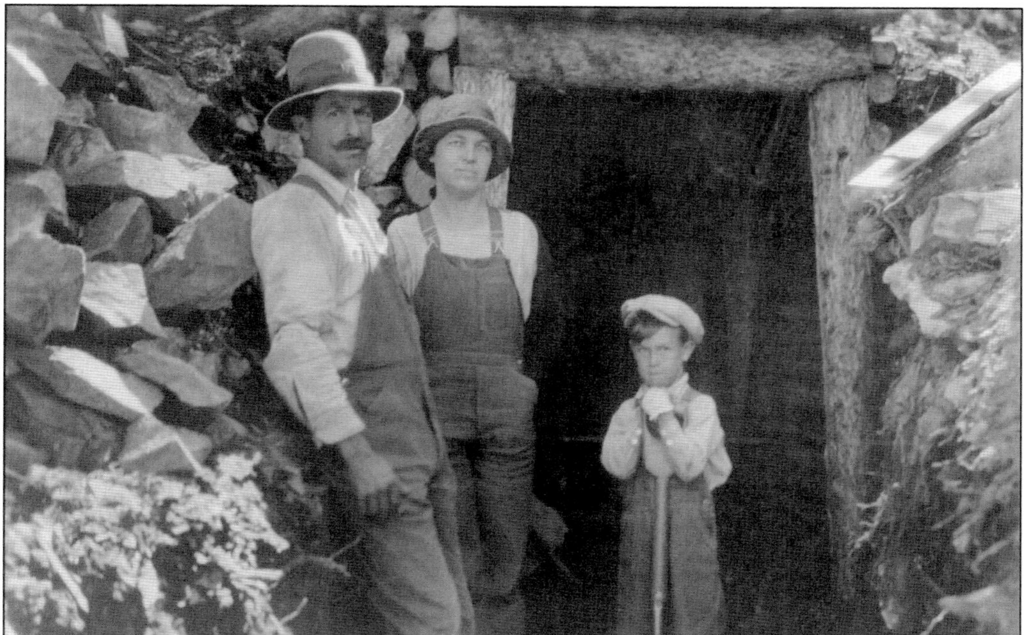

Written on the back of the photograph is "L.R. Melvin Martin Aug 1927 Rene and son Melvin some crew at the chrome tunnel after the entrance is all cleaned up. This was taken on our second trip up and the day was grand and oh the fun we did have." Chromium was discovered in the Goose Lake and Red Lodge area. (Courtesy of Mike Fleming.)

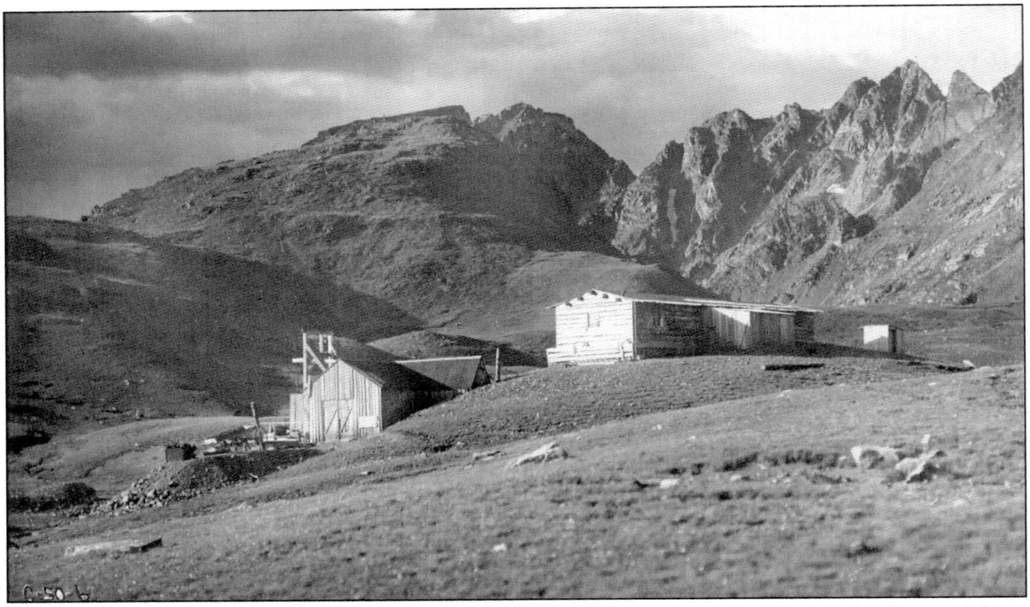

Copper King Mining, Milling, and Development Company was located 11 miles by road from Cooke City on the north end of Goose Lake at an elevation of 9,800 feet. Mount Zimmer is in the background. The road to Goose Lake was built by picks, shovels, and dynamite. The mine showed potential as a source of copper, gold, silver, and platinum. Copper is important in the age of electricity. Mine buildings, shafts, and tunnels were located above the timberline, so logs had to be hauled by horse and wagon for their construction.

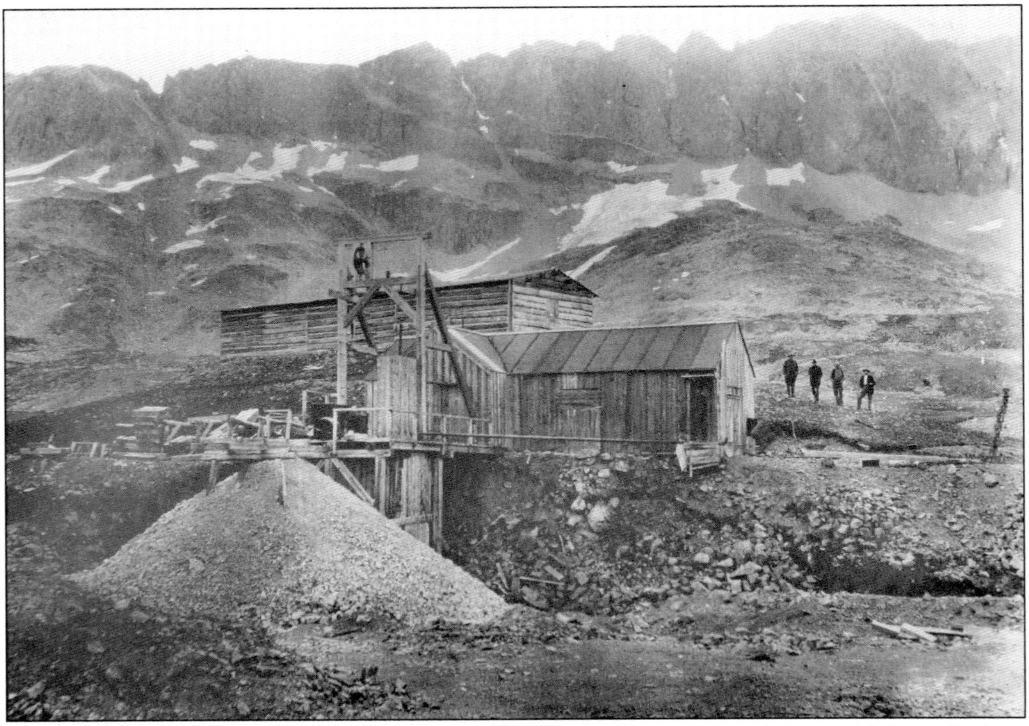

Discoverers of the rich deposits are, from left to right, Bert Holland, James Nelson, Robert Manderville, and F.C. Bryne.

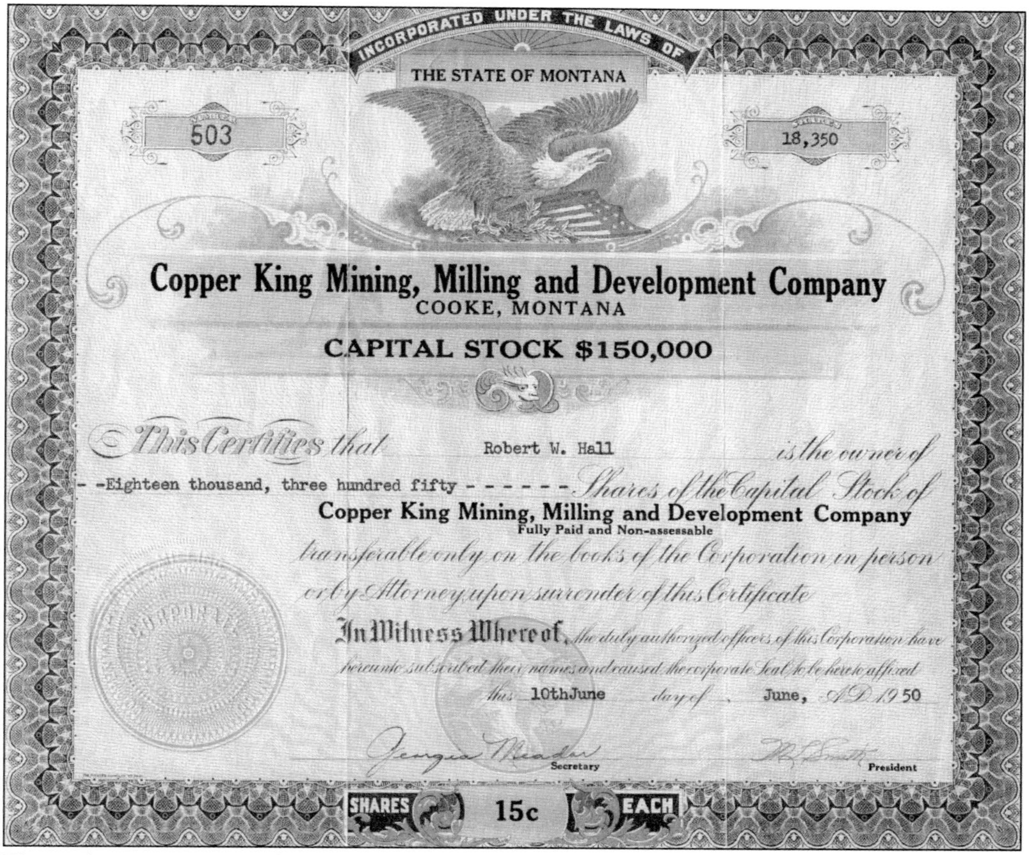
INCORPORATED UNDER THE LAWS OF

THE STATE OF MONTANA

503

18,350

Copper King Mining, Milling and Development Company

COOKE, MONTANA

CAPITAL STOCK $150,000

This Certifies that Robert W. Hall is the owner of

- -Eighteen thousand, three hundred fifty - - - - - - Shares of the Capital Stock of

Copper King Mining, Milling and Development Company

Fully Paid and Non-assessable

transferable only on the books of the Corporation in person or by Attorney upon surrender of this Certificate

In Witness Whereof, the duly authorized officers of this Corporation have hereunto subscribed their names and caused the corporate Seal to be hereto affixed this 10thJune day of June, A.D. 1950

Secretary

President

SHARES 15c EACH

This is the company's stock certificate. Headlines in the *Livingston Enterprise* read, "Million Dollar Deal Sold to Eastern Syndicate October 19, 1907." Investors made a substantial down payment, but the deal was never closed.

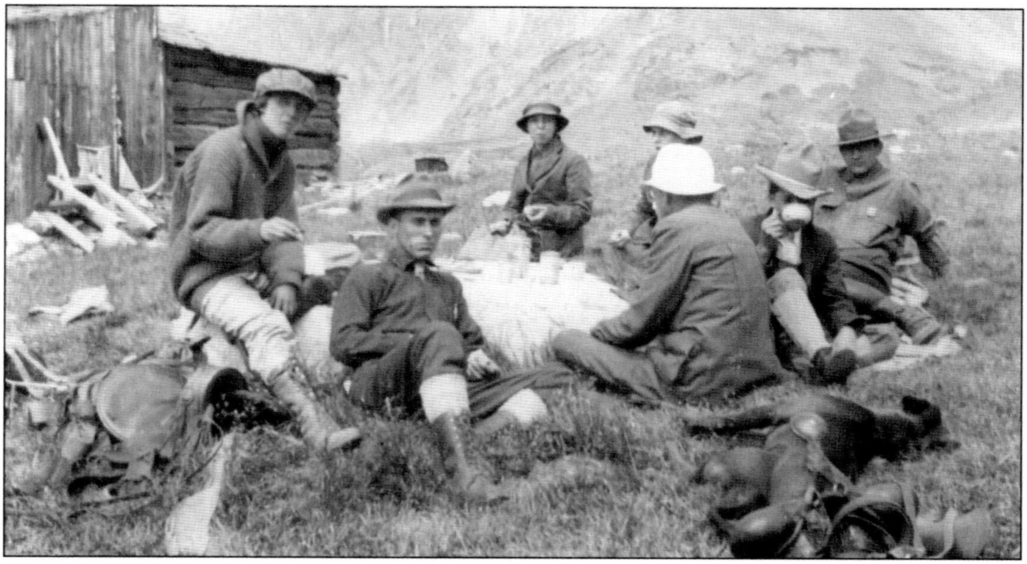

These workers are taking a break in front of Copper King Mining, Milling, and Development Company.

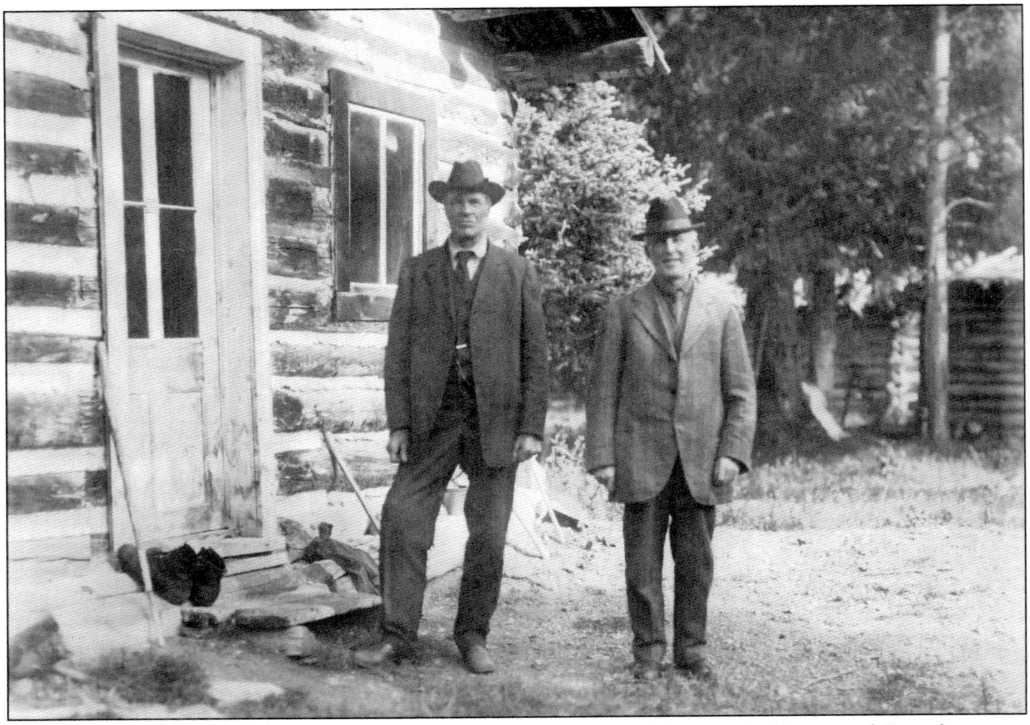

Frank Bryne (left), the one-time president of Copper King Mining, Milling, and Development Company, appears with his friend Tom Robbins.

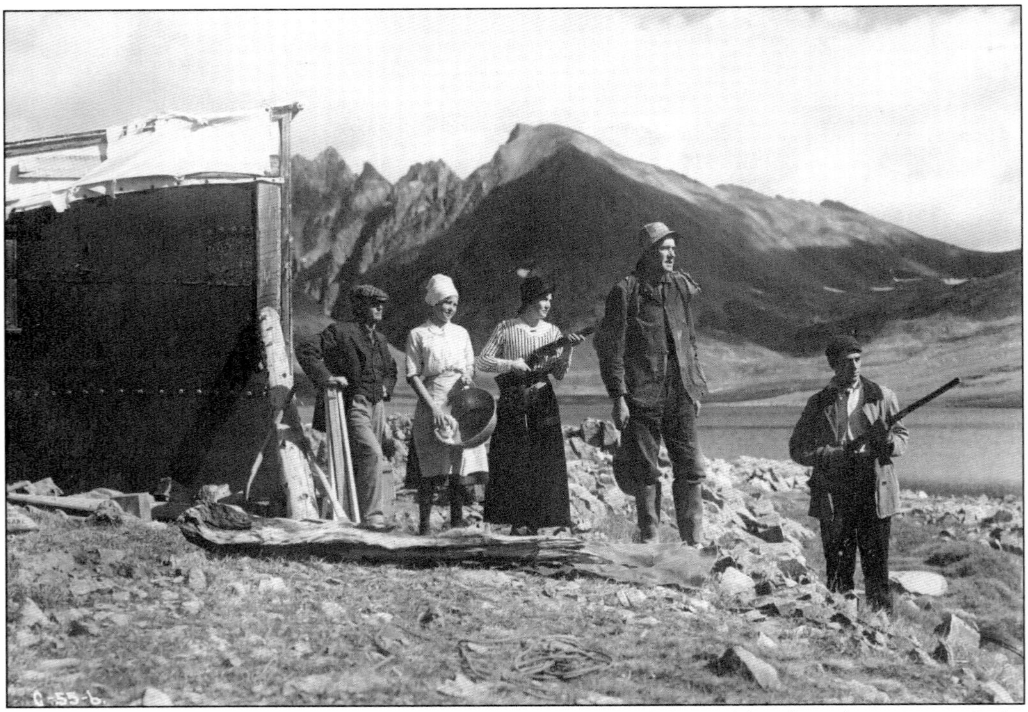

Posing for a photograph are O.B. "Bud" Hart, fourth from left, and his friends at Goose Lake near a crude dwelling.

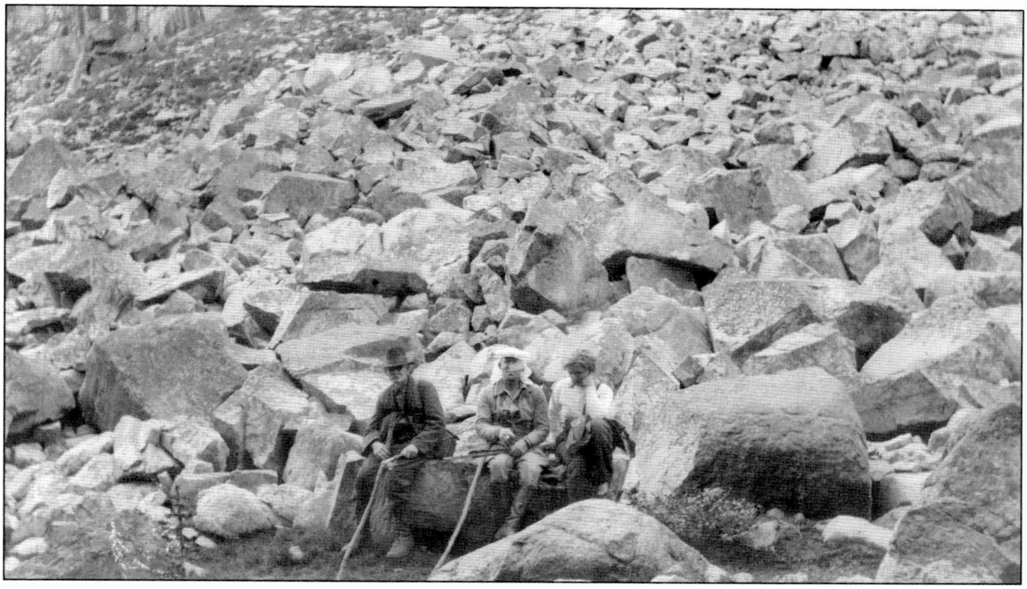

Granite glacial rocks provided a place for these hikers to rest while in the Goose Lake area; one gentleman even came prepared with his fishing pole.

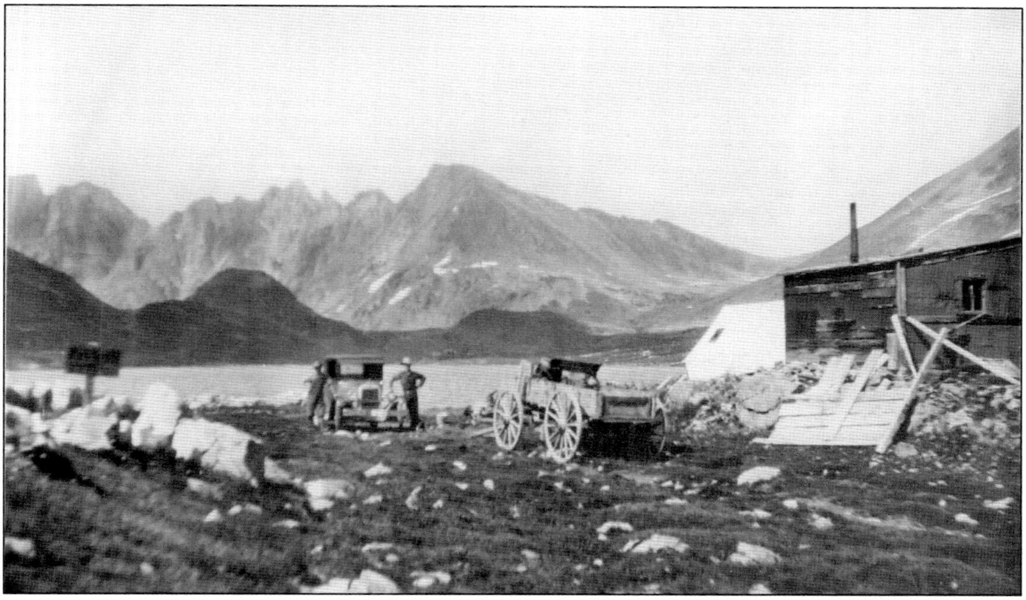

Written on the back of the picture is, "Goose Lake where Daddy's Copper projects are, Melvin Jr. age 18 and Daddy at his camp. This was the first and only car ever over the road." (Courtesy of Mike Fleming.)

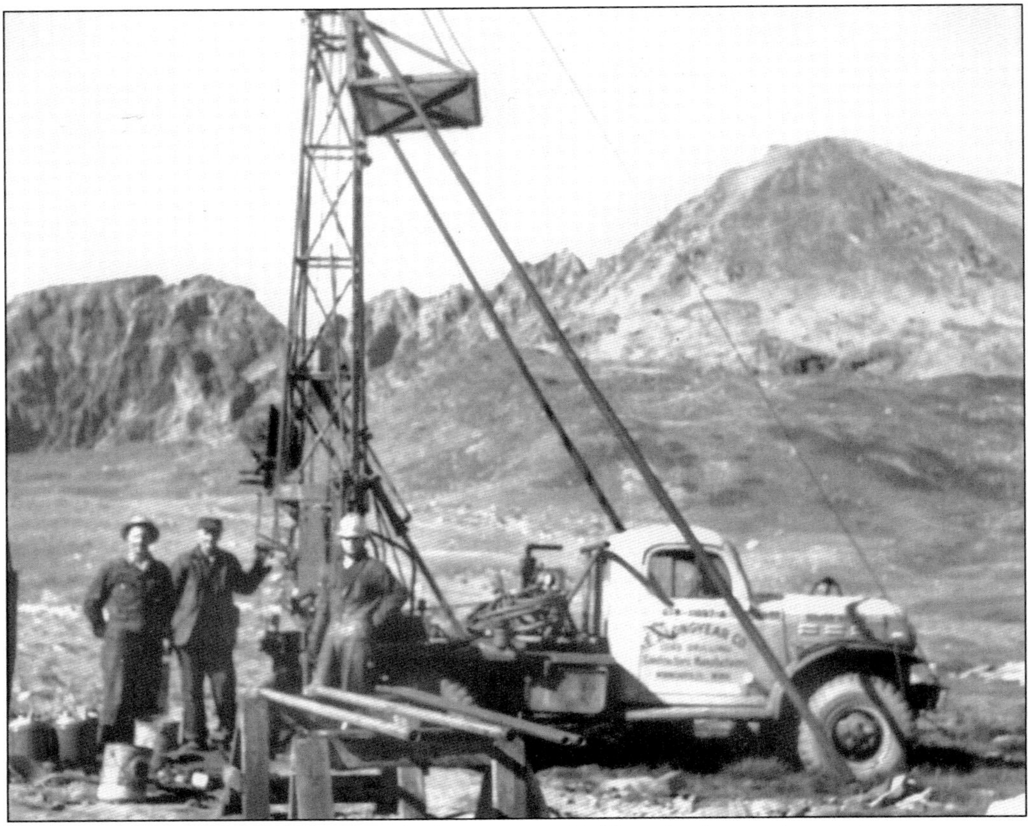

Explorations were done in the area by Kennecott, Gulf, and other major mining companies, resulting in drilling and geological studies. In 1988, Crown Butte Resources announced it they had made a major discovery of gold, silver, and copper at Henderson Mountain.

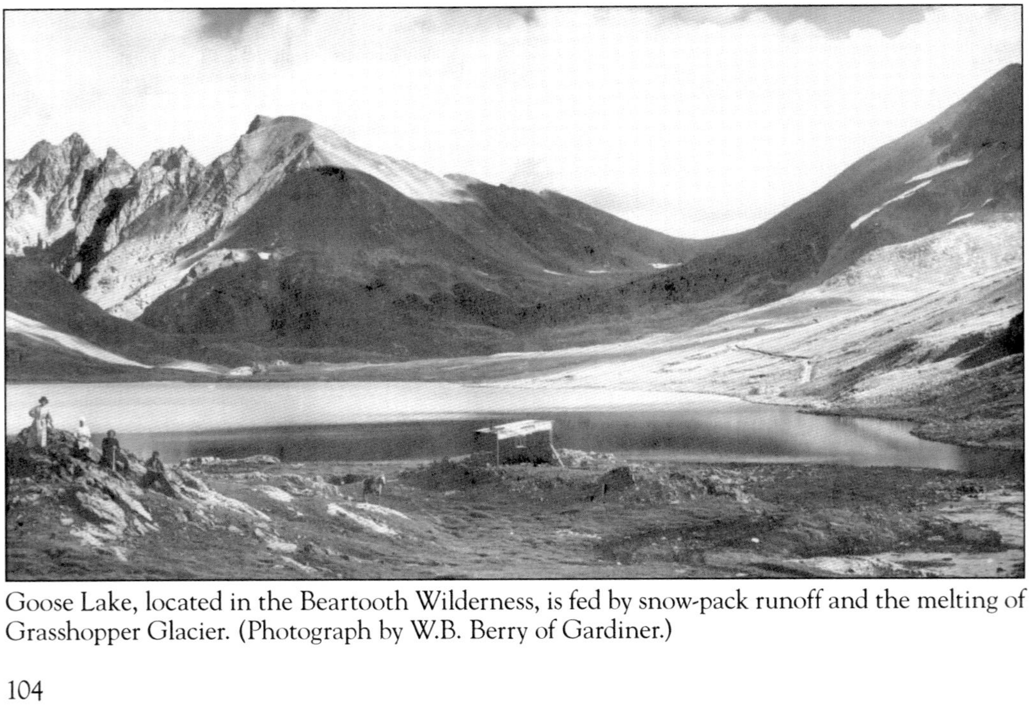

Goose Lake, located in the Beartooth Wilderness, is fed by snow-pack runoff and the melting of Grasshopper Glacier. (Photograph by W.B. Berry of Gardiner.)

Eight

Now and Then

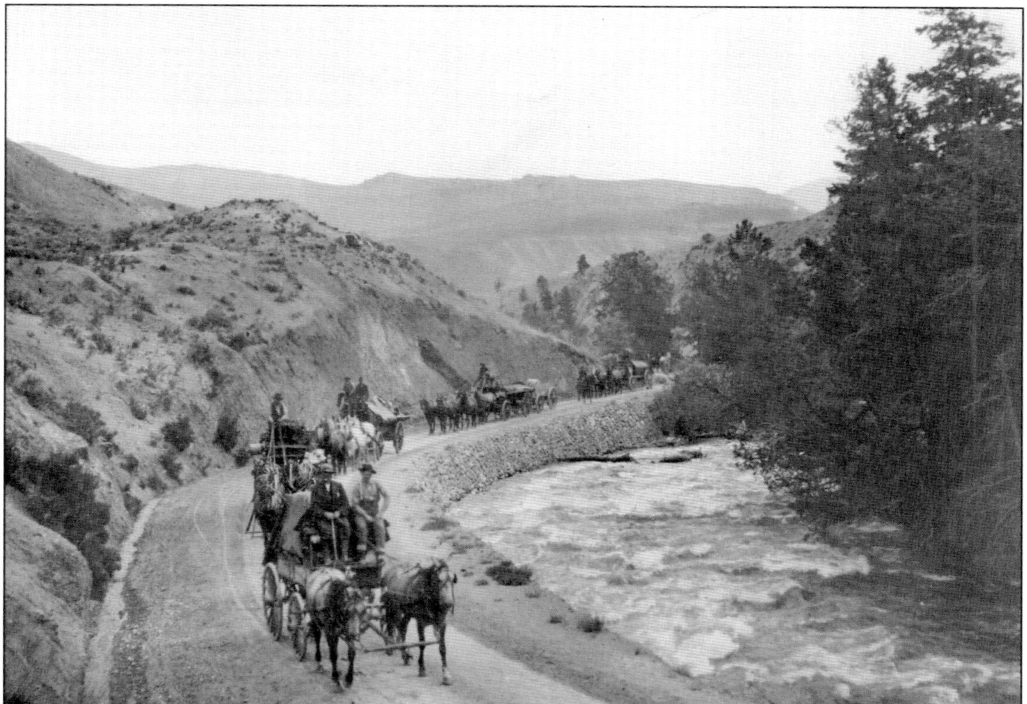

Crates and barrels of merchandise were shipped by rail to Gardiner, 60 miles from Cooke City. Taken between Gardiner and Mammoth, this 1915 photograph shows wagons freighting supplies through the park to Cooke City at a considerable rate less than a stagecoach. Geographically, Red Lodge was closer, but the highest and most rugged mountainous country separated them.

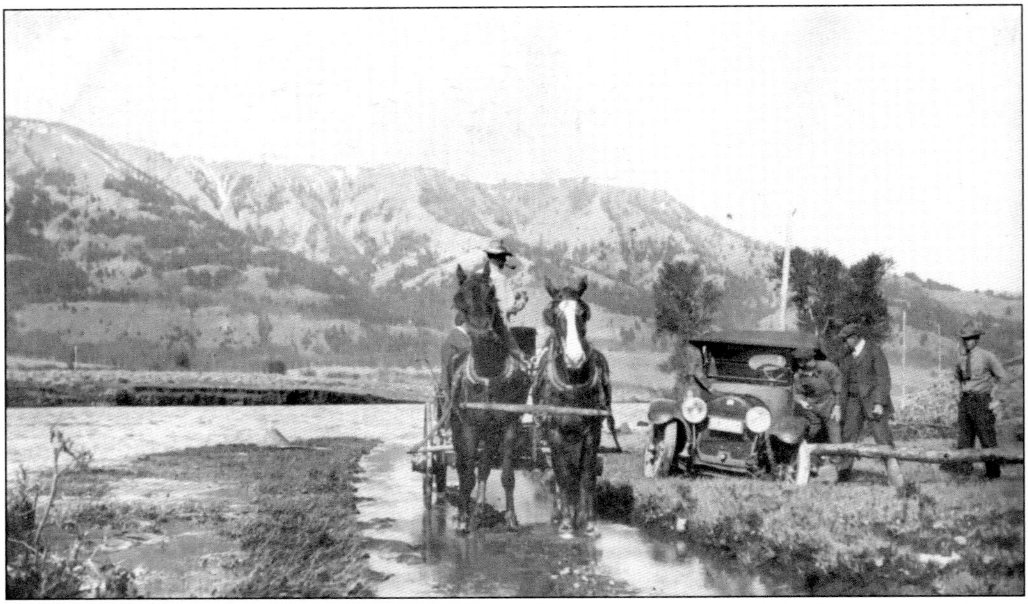

Ruts incised by wagons had a different wheelbase than those by automobiles. Drivers cursed mud that hindered travel.

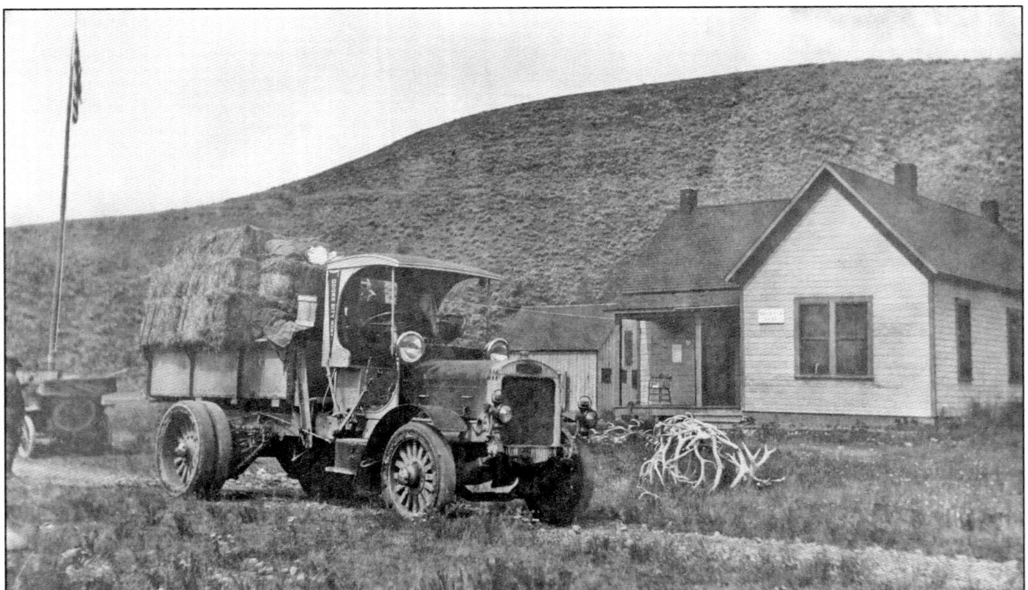

The auto truck in front of the Soda Butte Ranger Station is used to haul hay. The station is now known as the Buffalo Ranch. Hay was grown across the road near the river.

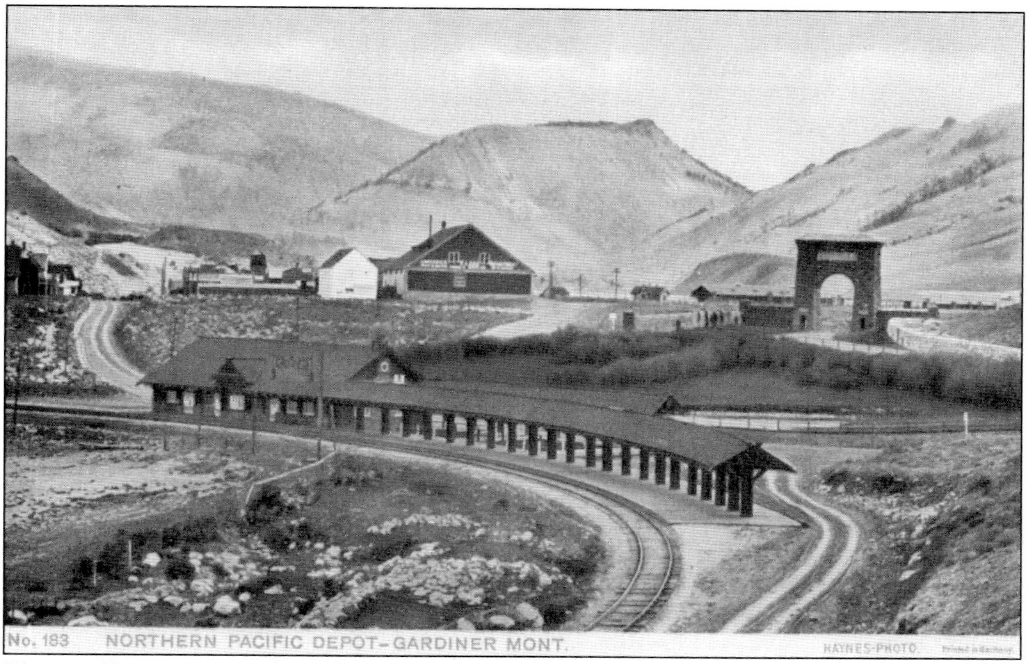

This is a Haynes postcard entitled "Northern Pacific Depot–Gardiner Mont." The Park Branch ended here. The famous arched entrance to America's first national park is on the right. The W.A. Hall Department Store stands in the background. It sold groceries, hay, oats, and just about anything else needed. It even had a tourist shop.

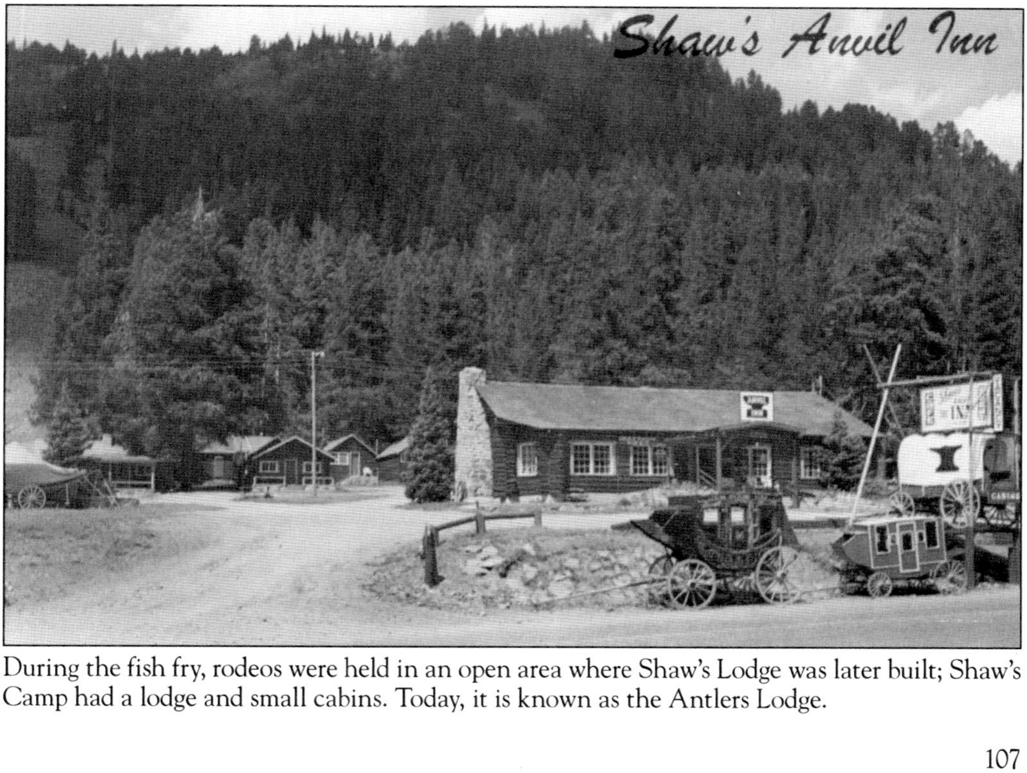

During the fish fry, rodeos were held in an open area where Shaw's Lodge was later built; Shaw's Camp had a lodge and small cabins. Today, it is known as the Antlers Lodge.

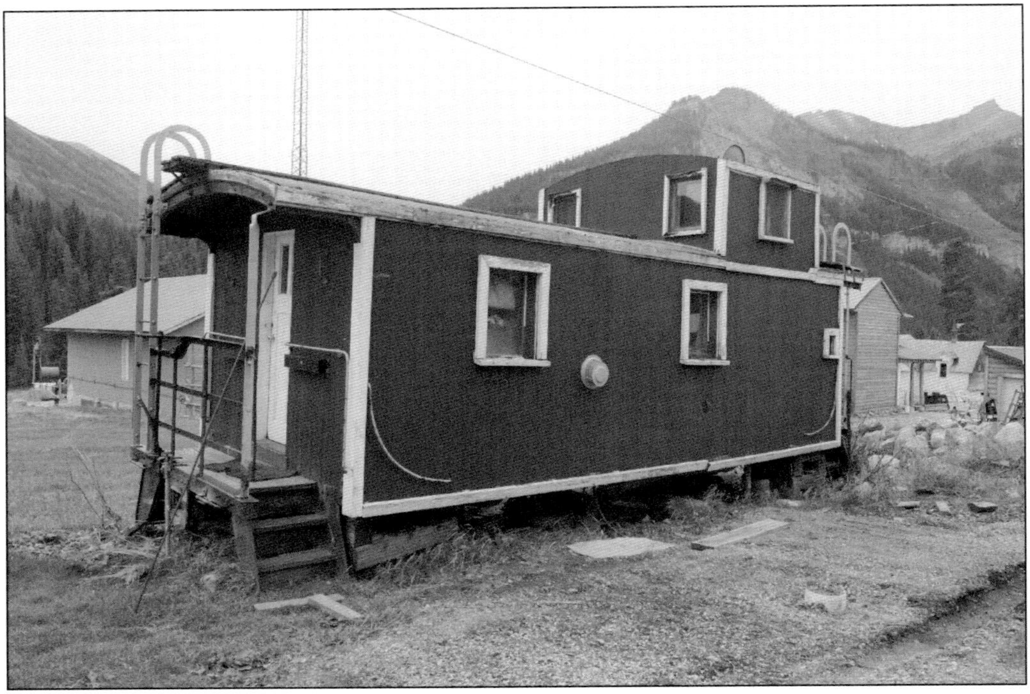

The residents of Cooke City someday hoped to hear the sound of the iron horse, but all they got was the tail end of a train, a caboose used as a nightly rental.

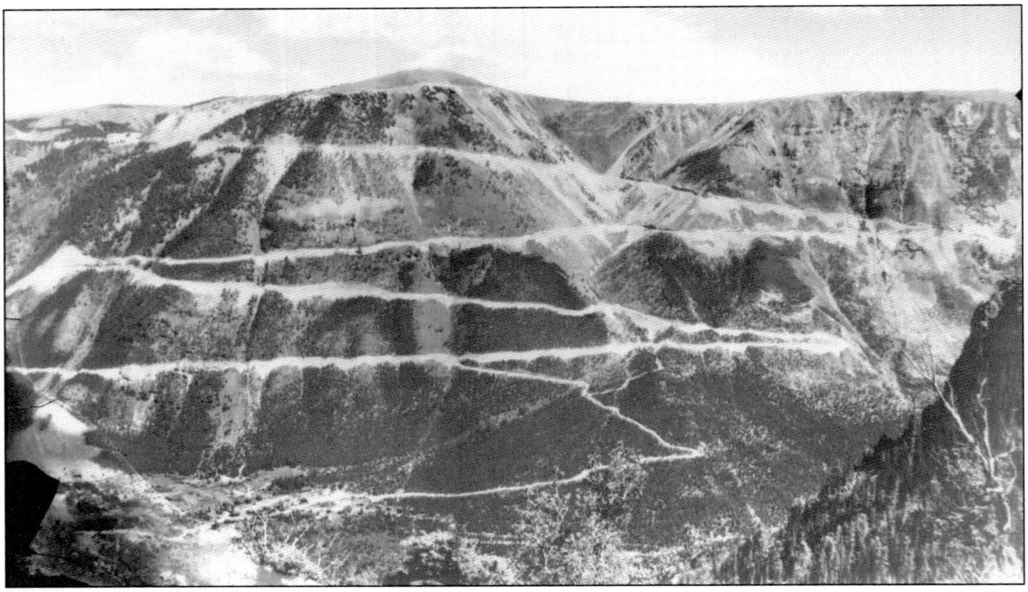

Cooke City had been handicapped by its inaccessibility, but this was remedied by the building of this magnificent new highway from Red Lodge through the Beartooth Mountains and down the Clark's Fork to Cooke City. This $3-million road began in 1932 and was completed four years later. It rises for 16 miles to the 10,940-foot pass and provides another entrance to Yellowstone National Park. The highway is plowed in the spring and is open as weather permits until fall. (Courtesy of Carbon County Historical Society and Donna Madson.)

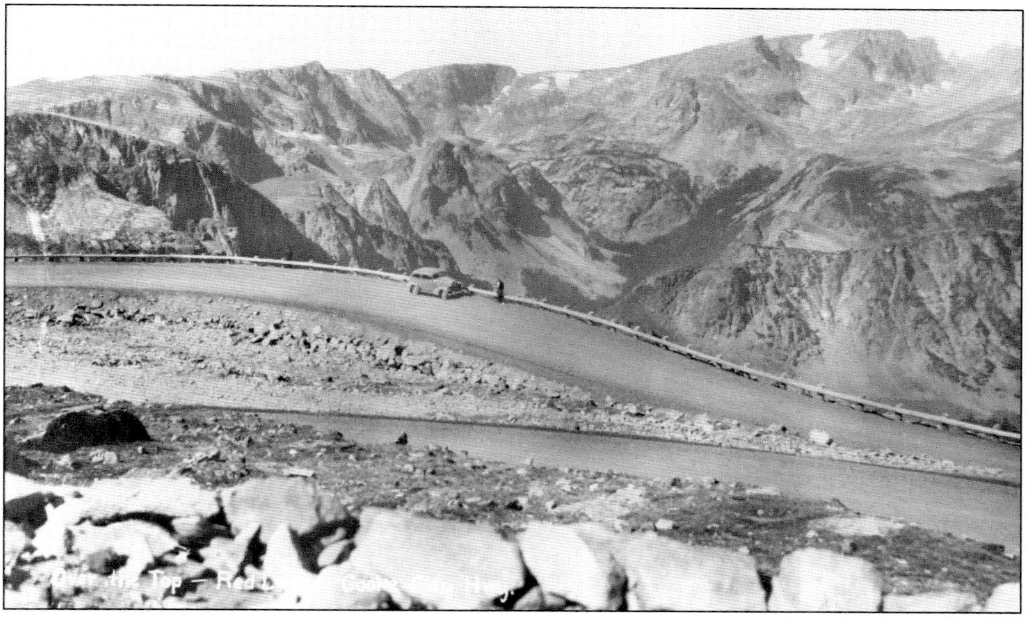

This old postcard, entitled "Over the Top," shows the fifth level at Cathedral Peak. From Red Lodge across Beartooth Pass to Yellowstone National Park is 68 miles. (Courtesy of Carbon County Historical Society, Donna Madson, Cooke City Montana Museum, and Dee Smith.)

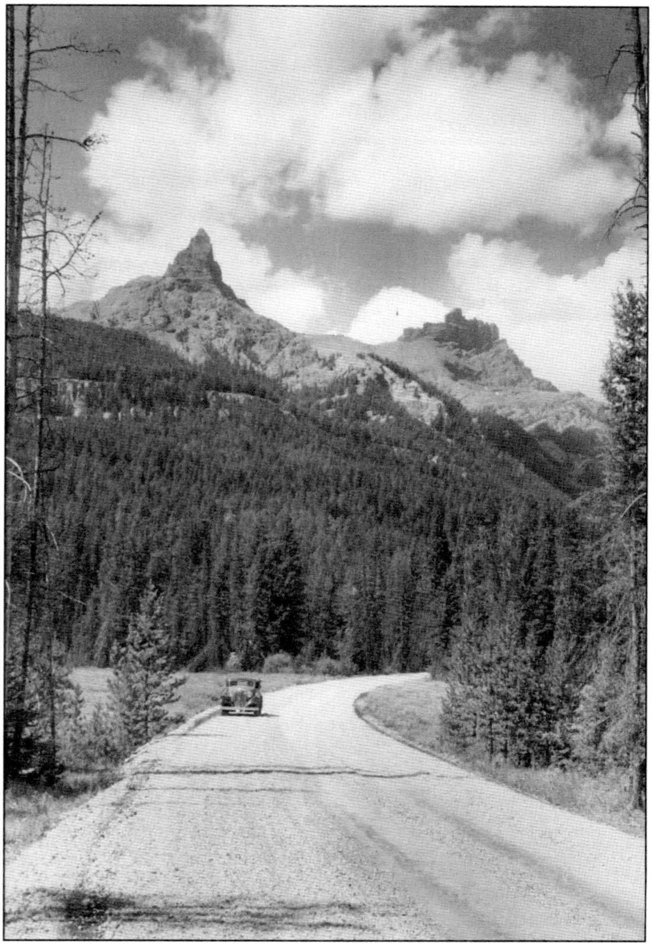

Index and Pilot tower above the road where this car is being driven in a Haynes photograph. When taking the Chief Joseph Scenic Highway to Cody, Wyoming, 15 spectacular mountain peaks can be viewed from the 8,000-foot summit of Dead Indian Hill. Through this pass, great herds of game traveled from the mountains to the plains. In 1877, Chief Joseph led his tribe, the Nez Perce, over this pass pursued by US soldiers. This road joins the Beartooth Highway 12 miles east of Cooke City. This area hosts many campgrounds.

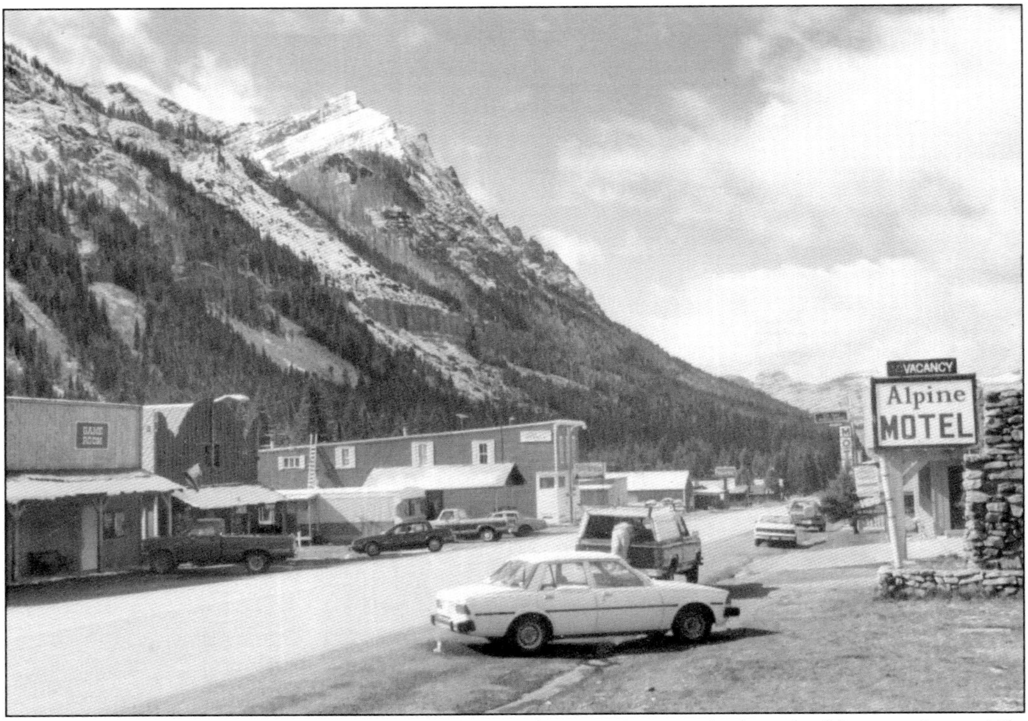

This is Main Street looking west in 1992. The Cooke City Store is the long building in the middle of the photograph and was painted red, as it is today. (Courtesy of Warren McGee.)

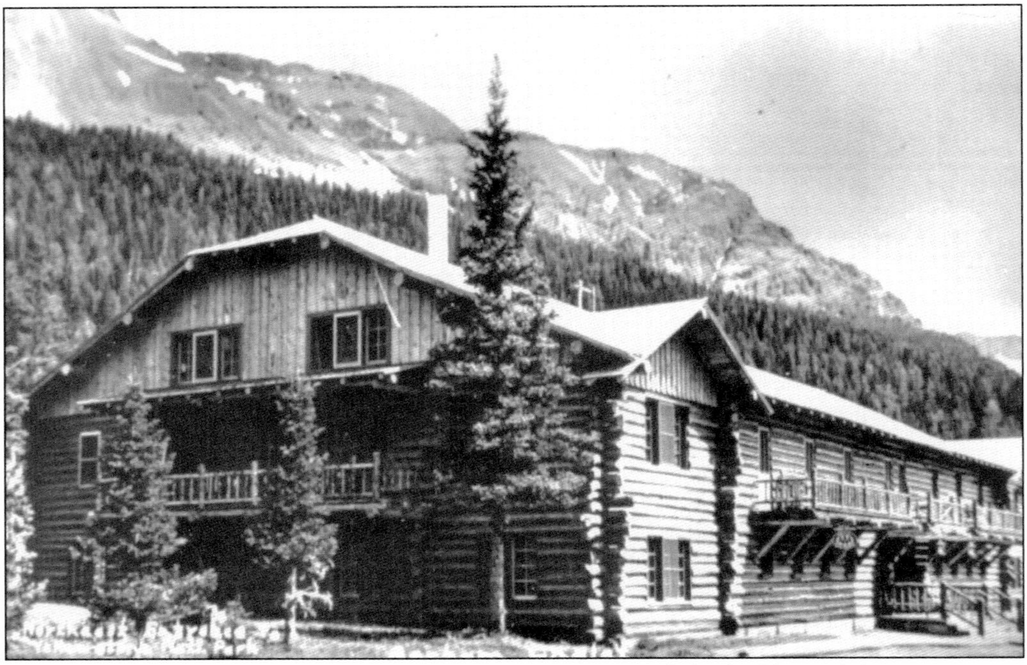
On September 8, 1932, Park County commissioners approved the building of the rustic resort town of Silver Gate. It is located one mile from the northwest entrance of Yellowstone National Park and three miles from Cooke City and has been called the Little Switzerland of Montana. This is the Gorham Chalet, now known as the Range Riders.

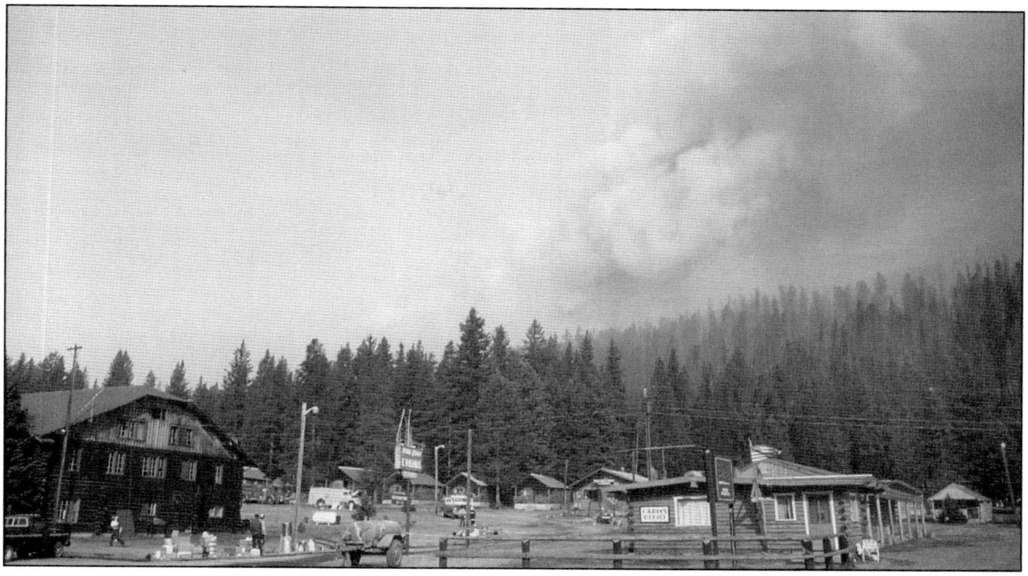

In 1988, after a summer of extreme heat, lightning started fires in Yellowstone National Park. As the fire sped toward Silver Gate, these smoke plumes were captured by Jim Peaco. Fire raged in the mountains just outside of Cooke City, which became a firebase. Backfires were set to burn up fuel on the path of the fire to create a firebreak but got out of control. Courageous firefighters battled the flames to control the fire and save human life and property in the Silver Gate–Cooke City area. The perimeter of the area burned, and some buildings and homes were lost in the outlying area. On the sign on the edge of town, someone added the letter *d* to the end of Cooke. (Courtesy of Yellowstone National Park, Jim Peaco.)

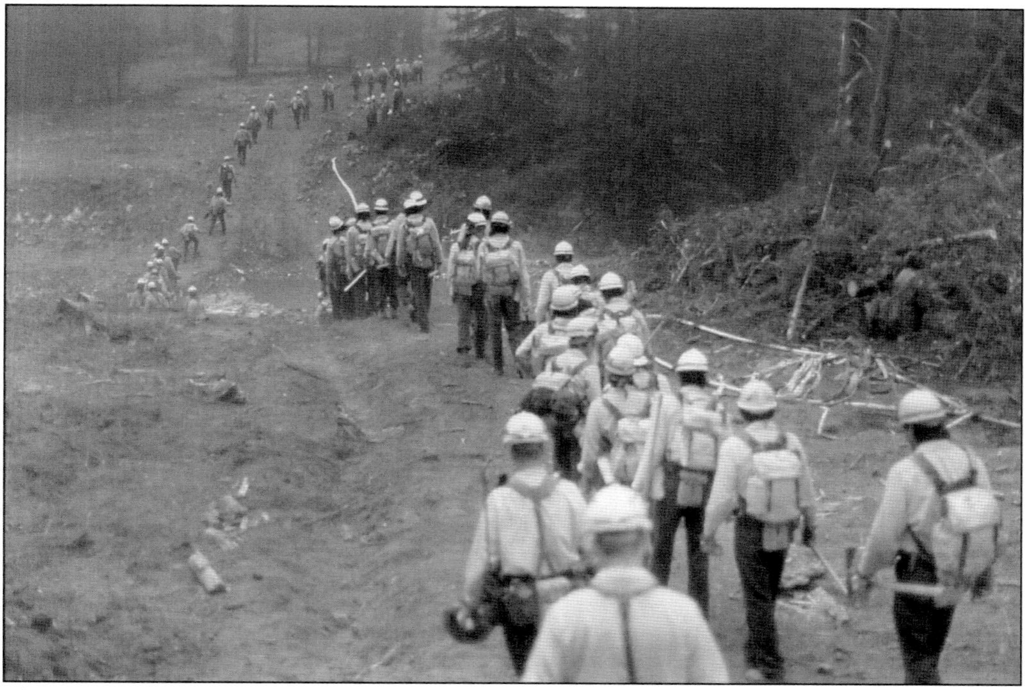

This image by Jim Peaco shows the army of firefighters that was called in to battle the blazes. (Courtesy of Yellowstone National Park.)

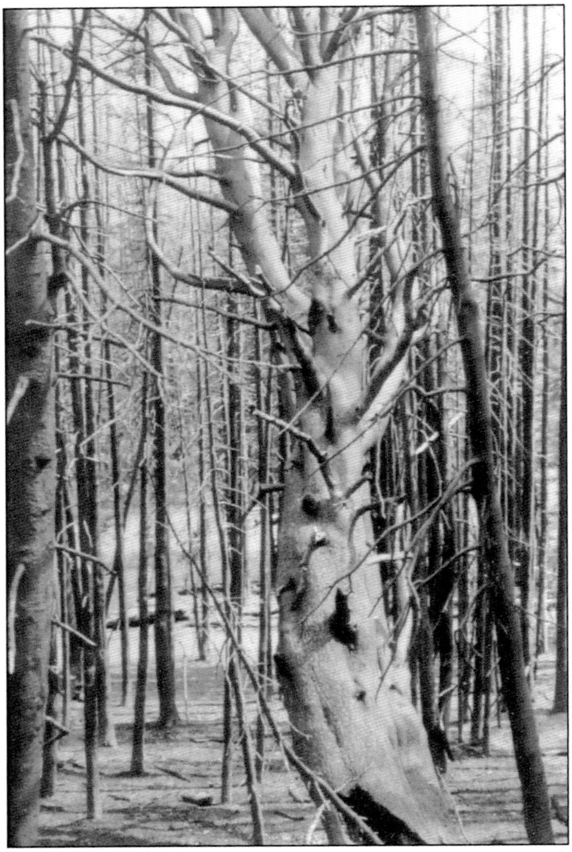

Charred trees and grasses remained from the fire. As quickly as the fire came and went, the forests and grasses began to rejuvenate. The next spring, the fire grass was spectacular. The let-burn policy was questioned by many, but it was established that fires are necessary to maintain the overall health of the Yellowstone ecosystem.

Over the years, more than 100,000 tons of ore was in dumps. Almost every camp had dumps containing important minerals that were not recovered because of the expense. With newer mining methods, waste dumps yielded their hidden wealth.

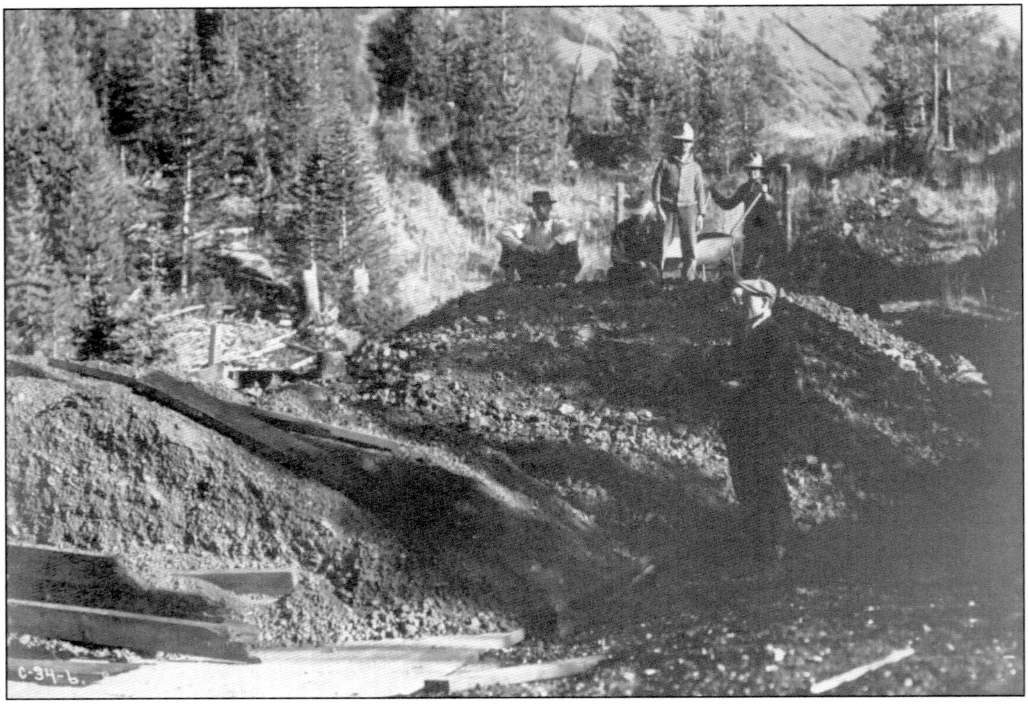

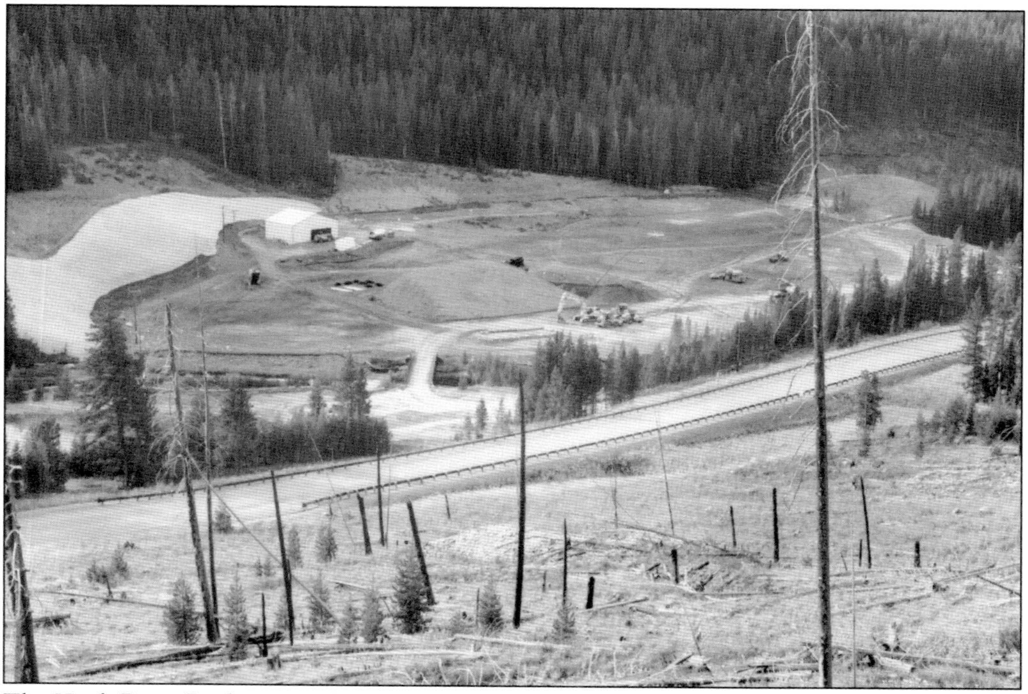

The Knife River Reclamation Project is to restore devastation caused by historic mining in the area of the McLaren Mine east of Cooke City. Some of Soda Butte Creek will be restored to its original bed. The trees in the foreground remain from the 1988 fires.

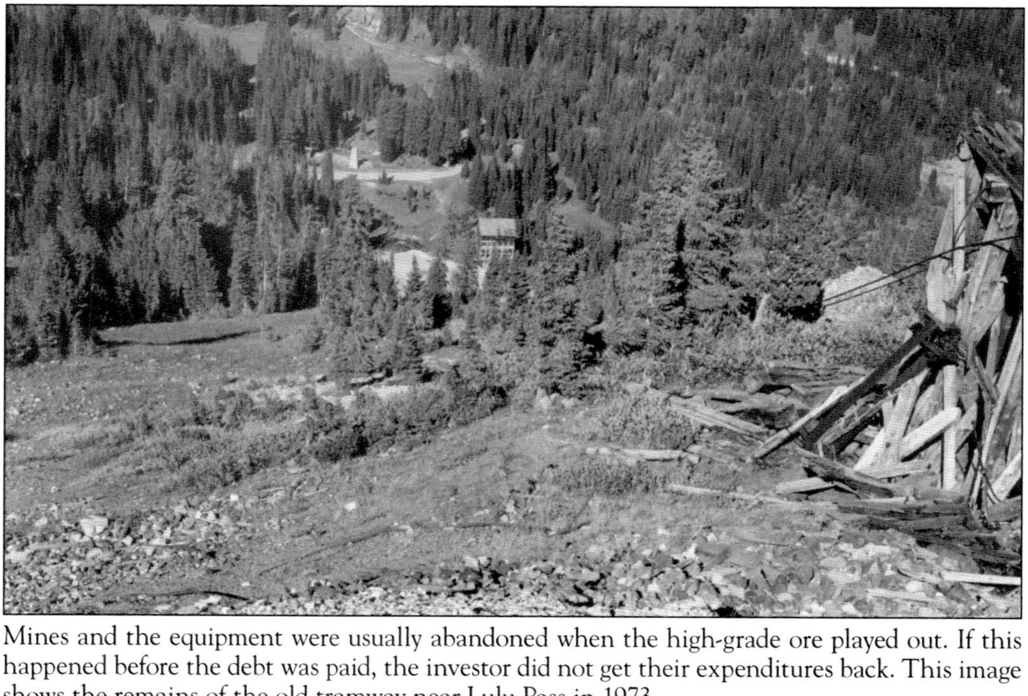

Mines and the equipment were usually abandoned when the high-grade ore played out. If this happened before the debt was paid, the investor did not get their expenditures back. This image shows the remains of the old tramway near Lulu Pass in 1973.

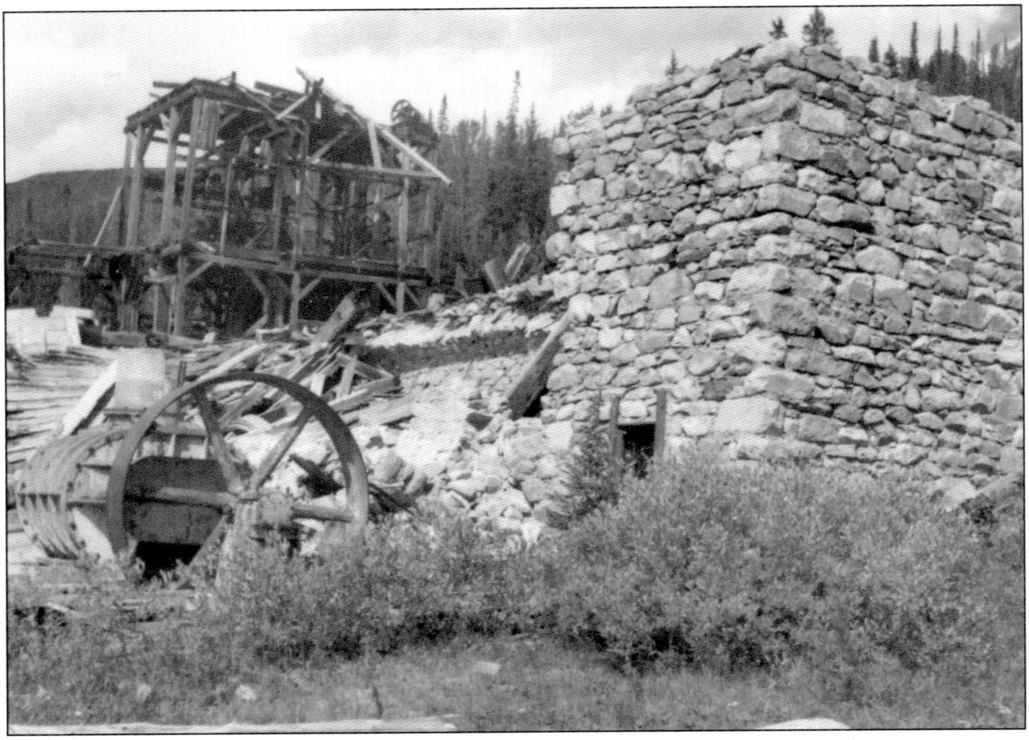

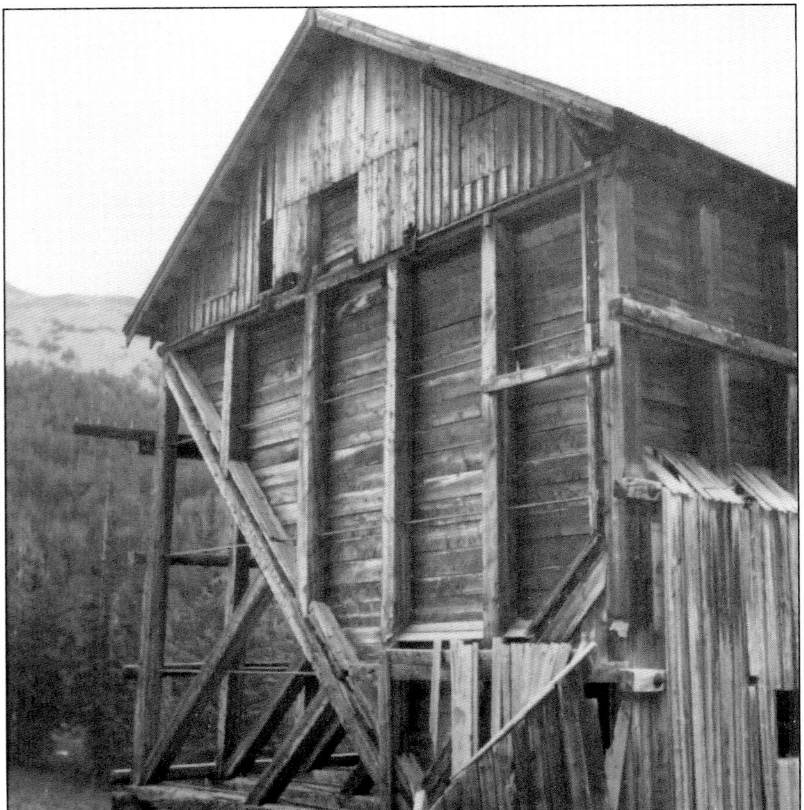

This stone foundation and smelter equipment were also remaining at that time. Fire, decay, heavy snow of winter, and pilfering took a toll on the remaining historic structures.

Ore from the Homestake went through a crusher, onto a conveyor belt, and was loaded into a tram system and sent down Fisher Creek to Western Smelting & Power Company's smelter.

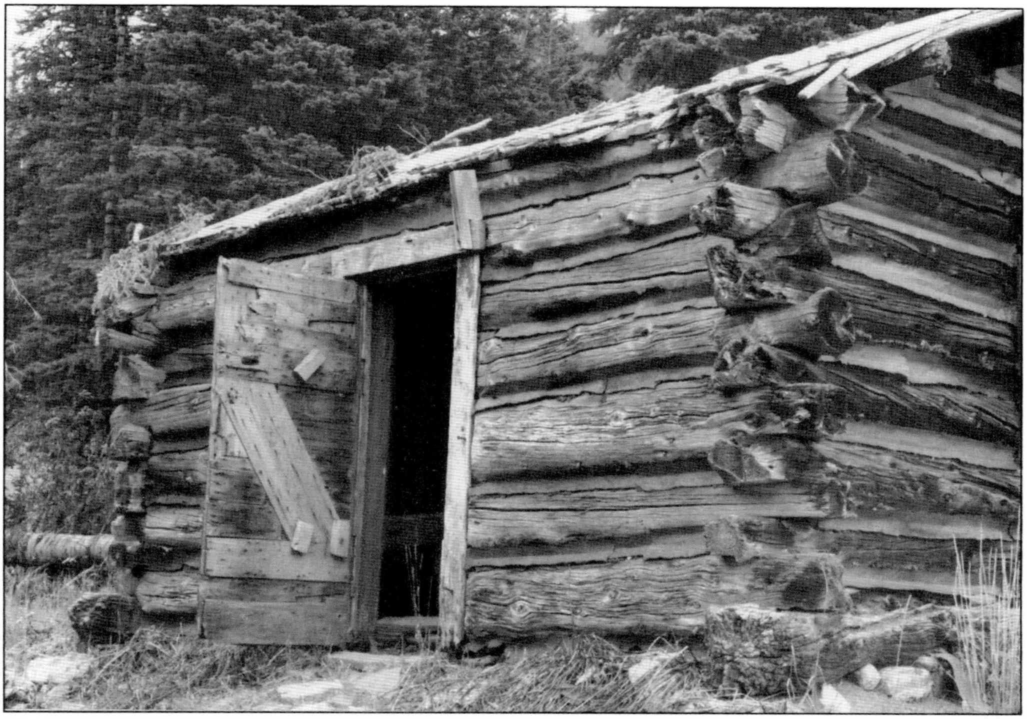

Horn Miller lived close to nature. Journals found in his cabin attest to the many mountain bikers, skiers, hikers, and others who have stayed in the cabin. The cabin is below Crown Butte and has been well preserved and protected by its inhabitants.

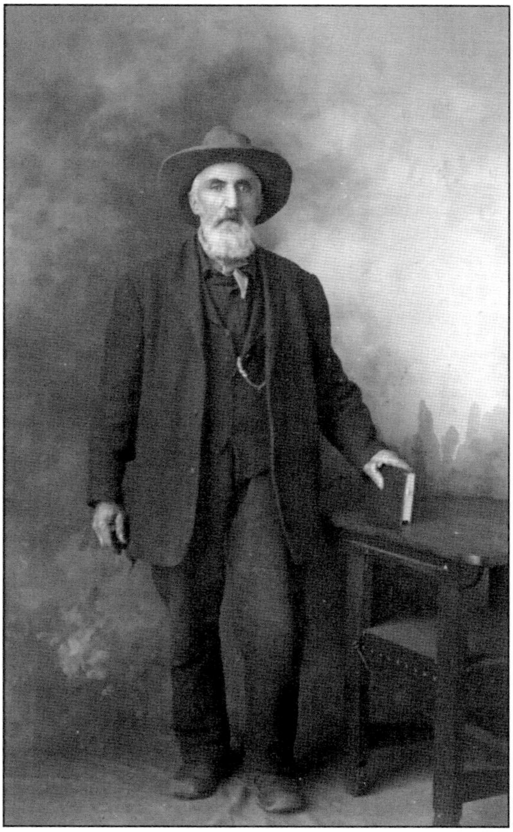

Horn Miller was said to have been nearly six feet tall and strong as a moose. In his obituary, he was described as a man of sterling character, a man without enemies of any kind, and a citizen who had a kind word for everyone. He was buried in the cemetery under the largest tree, facing east as he requested so he could see the sun come up every morning.

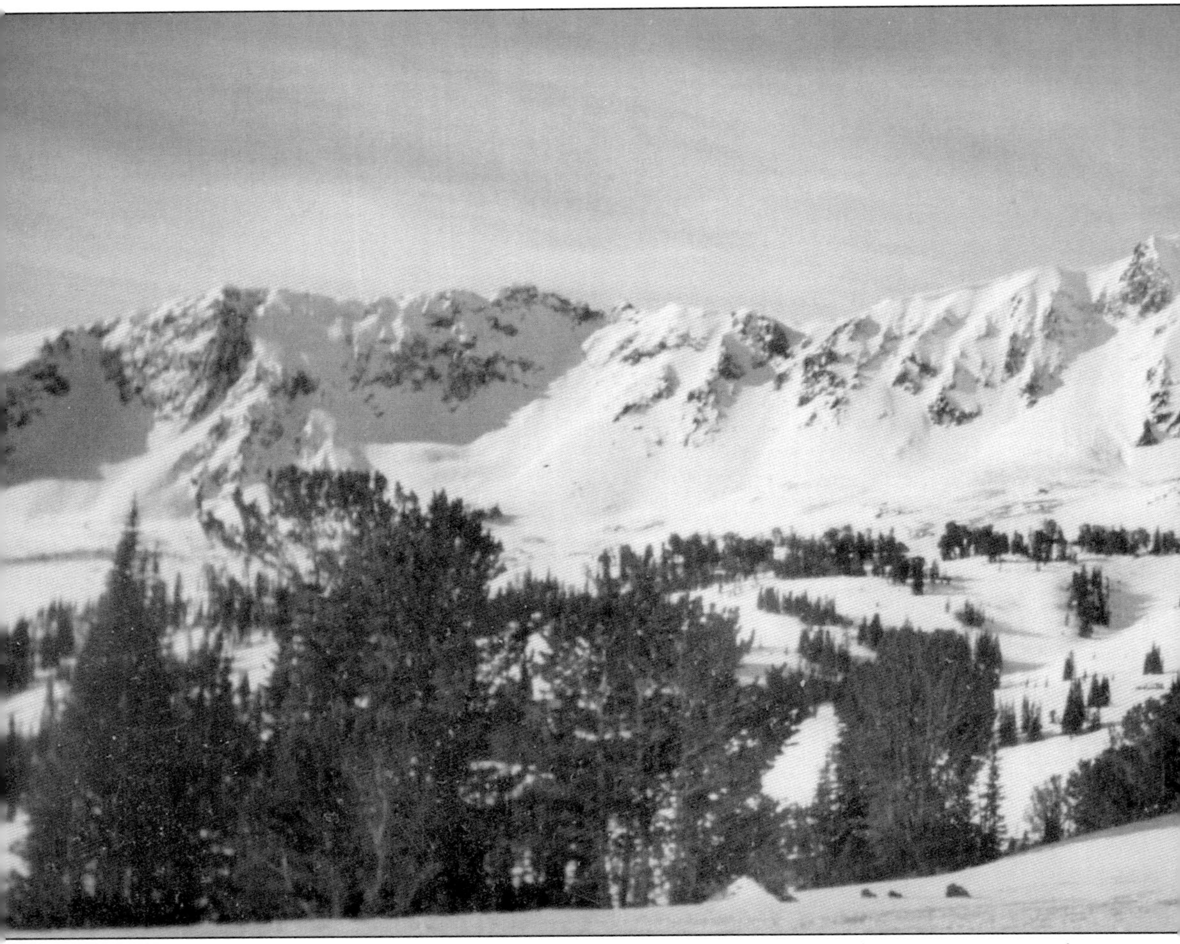

Cooke City areas offer winter recreational activities, including snowmobiling, cross-country skiing,

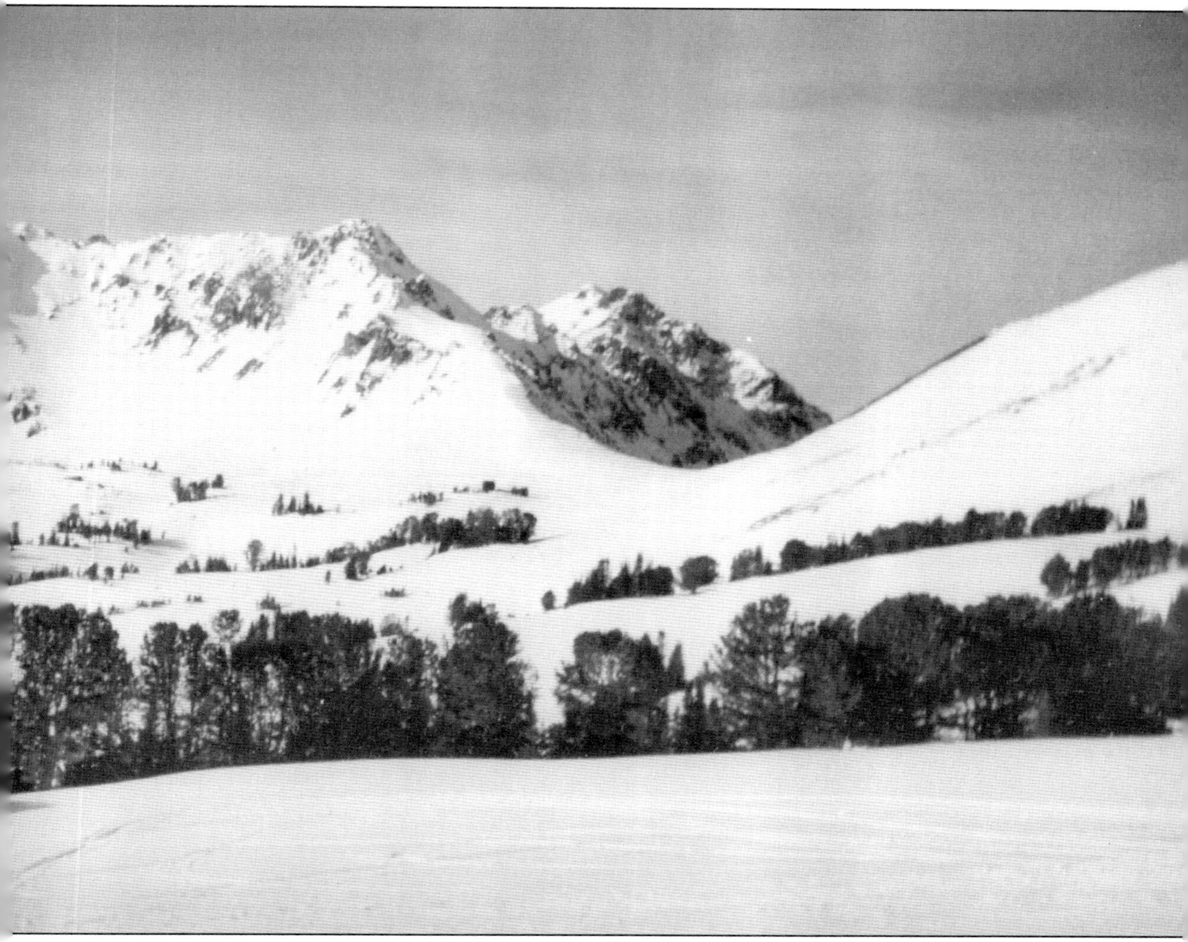

snowshoeing, hunting, and many other outdoor activities.

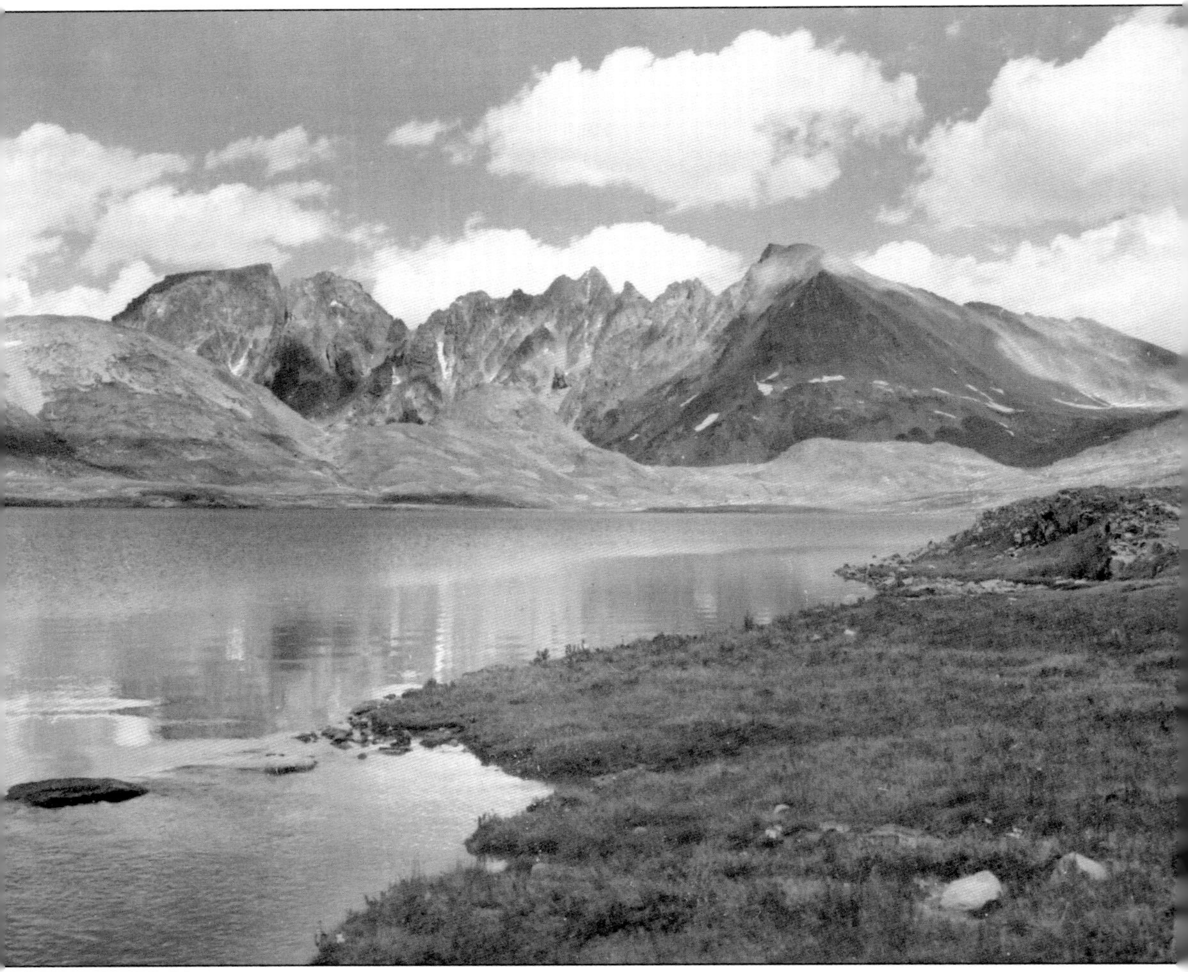

In the summer, fishing, mountain biking, four wheeling, Jeeping, backpacking, camping, picnicking, animal watching, sightseeing, and photography were popular activities. Also, Yellowstone National Park is nearby.

The De Loris Falls was named after De Loris Cole, a young lady from Cooke City.

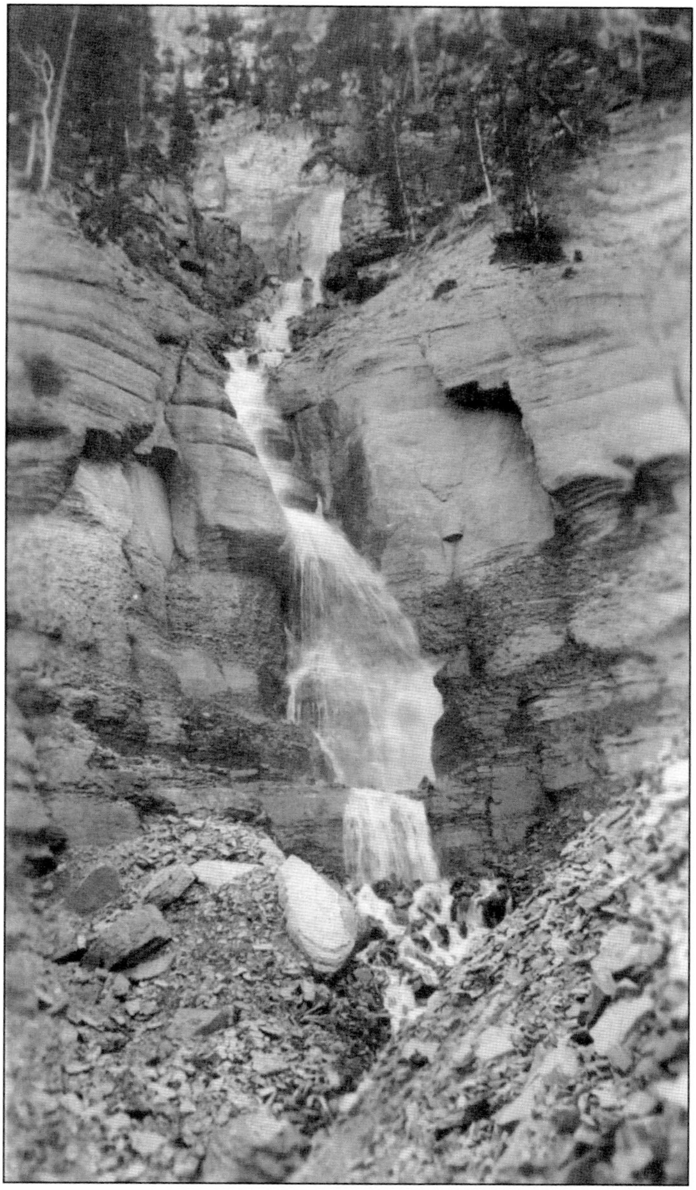

A.B. Henderson had written in his journal of mountains rising like walls thousands of feet high. Republic Mountain, with its rocky crest, hems in the valley on the southwest. Soda Butte Creek meanders below it.

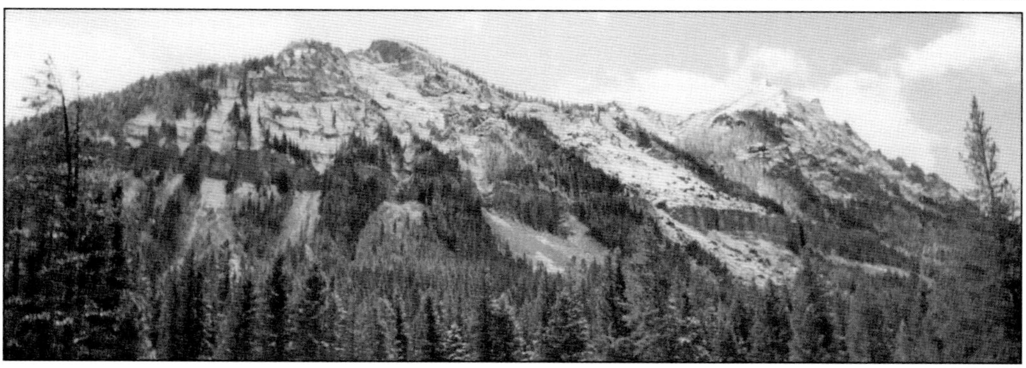

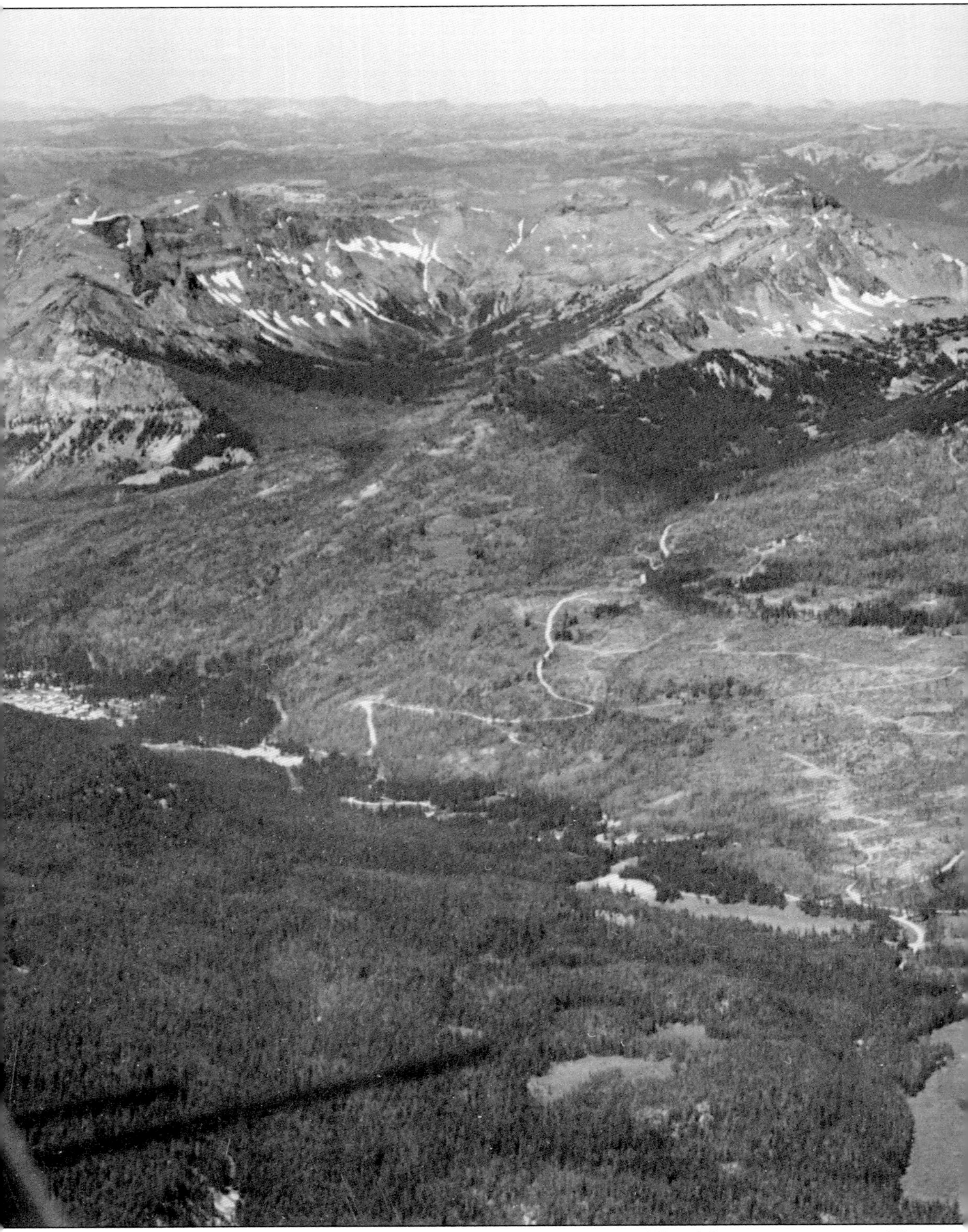

Cooke City is in the lower left and Cooke Pass is in the lower middle of this panoramic view. The prominent topography from left to right is Mineral Mountain, Miller Mountain, Bull of the Woods,

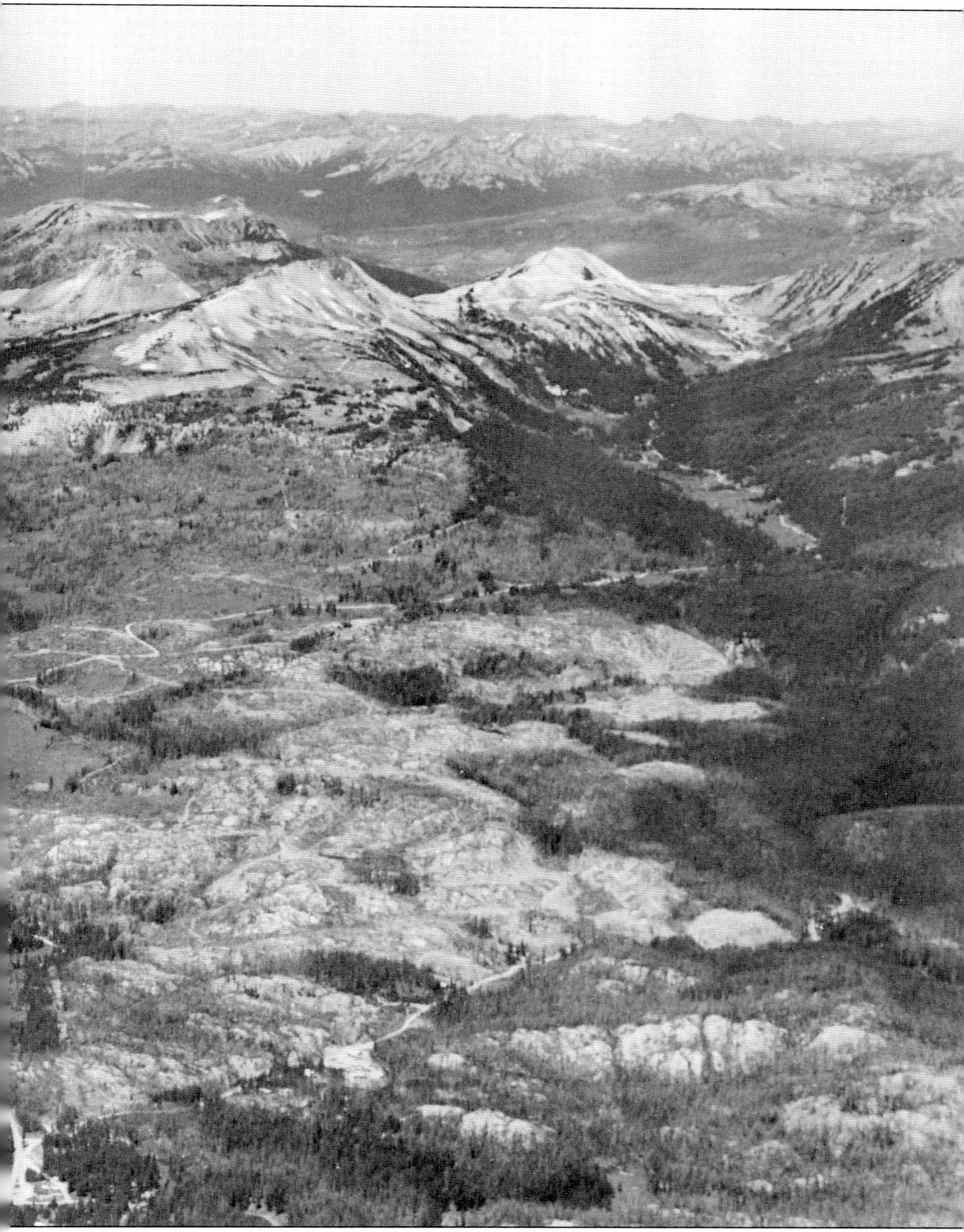

Crown Butte, Daisy Pass, Henderson Mountain, and Lulu Pass. The gray area in the middle of the photograph shows the extent of the 1988 fires. (Courtesy of Crown Butte Mining Company.)

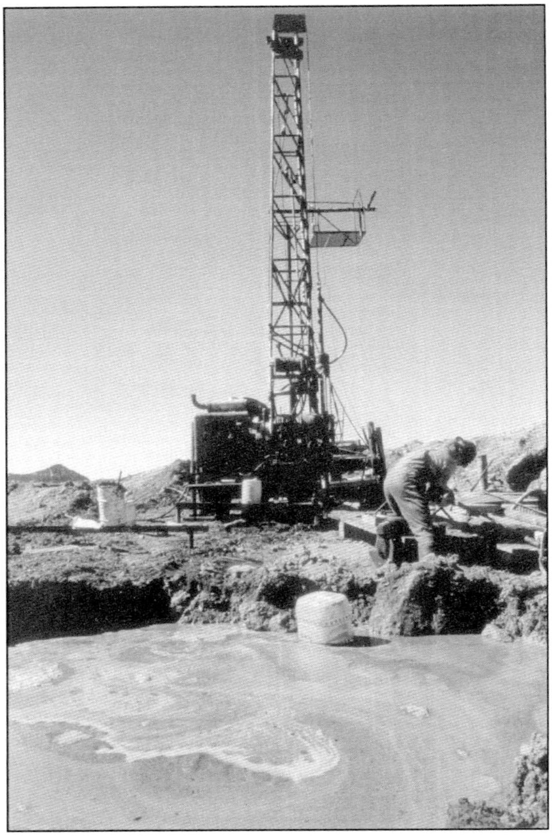

In 1989, Crown Butte Mines proposed a large gold mine less than three miles from Yellowstone National Park on Henderson Mountain. Gold mining had started in the 1870s on Henderson, which was named after Bart Henderson. This is a company drill.

The rich discovery was reported to be worth from $800 million to nearly $1 billion in gold, silver, and copper. This core sample is .027 percent gold, 1.01 percent silver, and .808 percent copper.

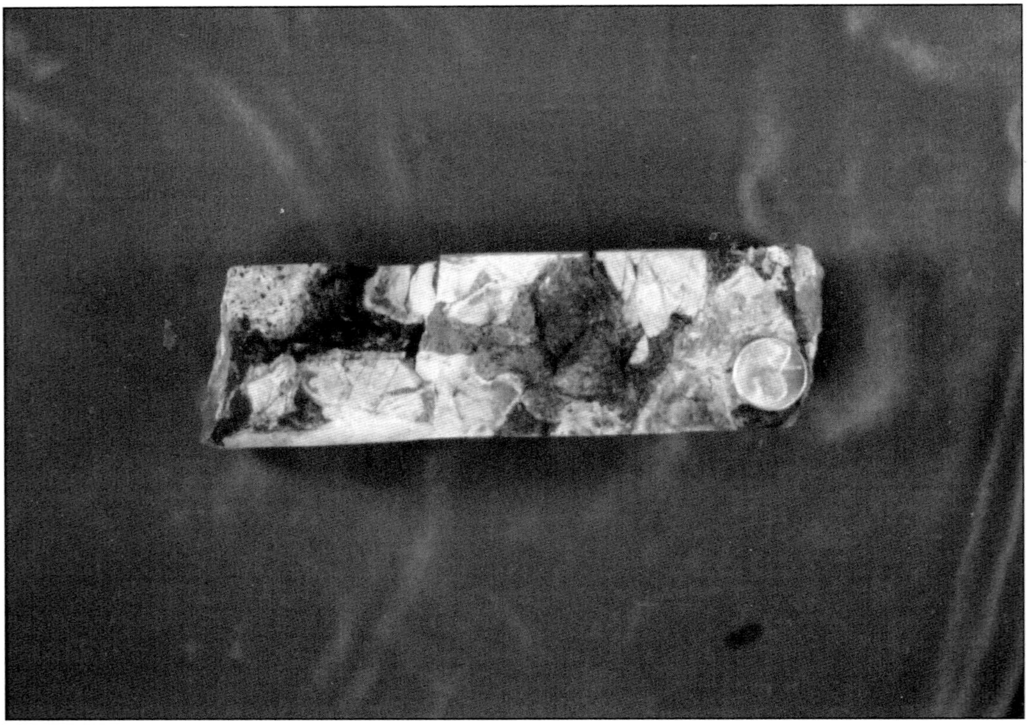

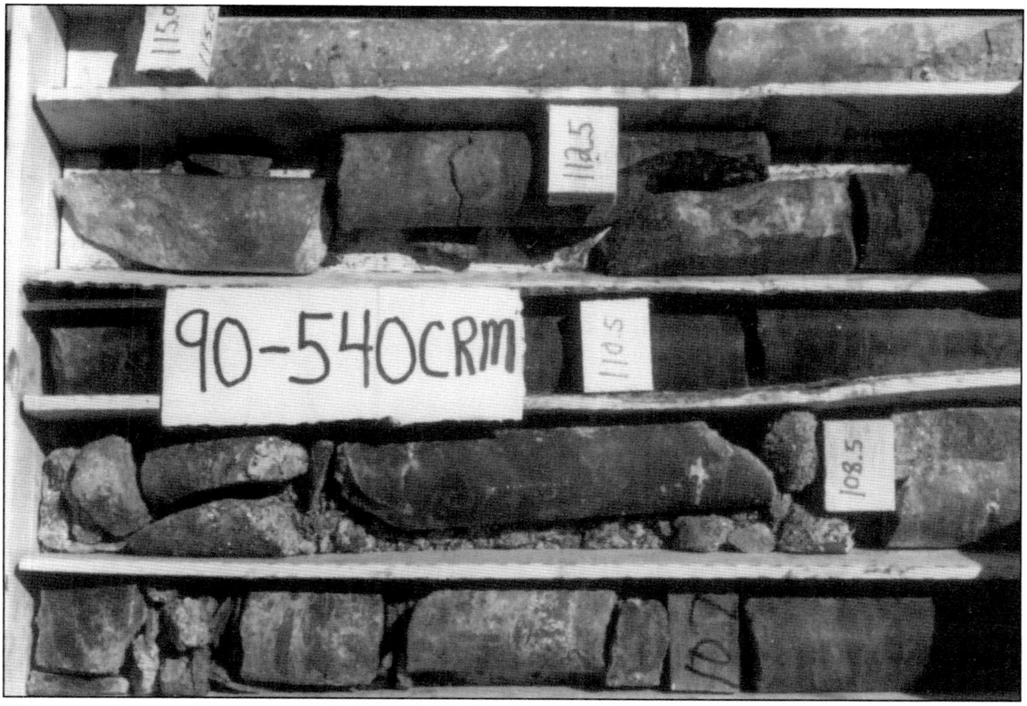

To access the ore deposits, six drill rigs drilled 24 hours a day. Boxes of core samples are numbered identifying where the core was taken.

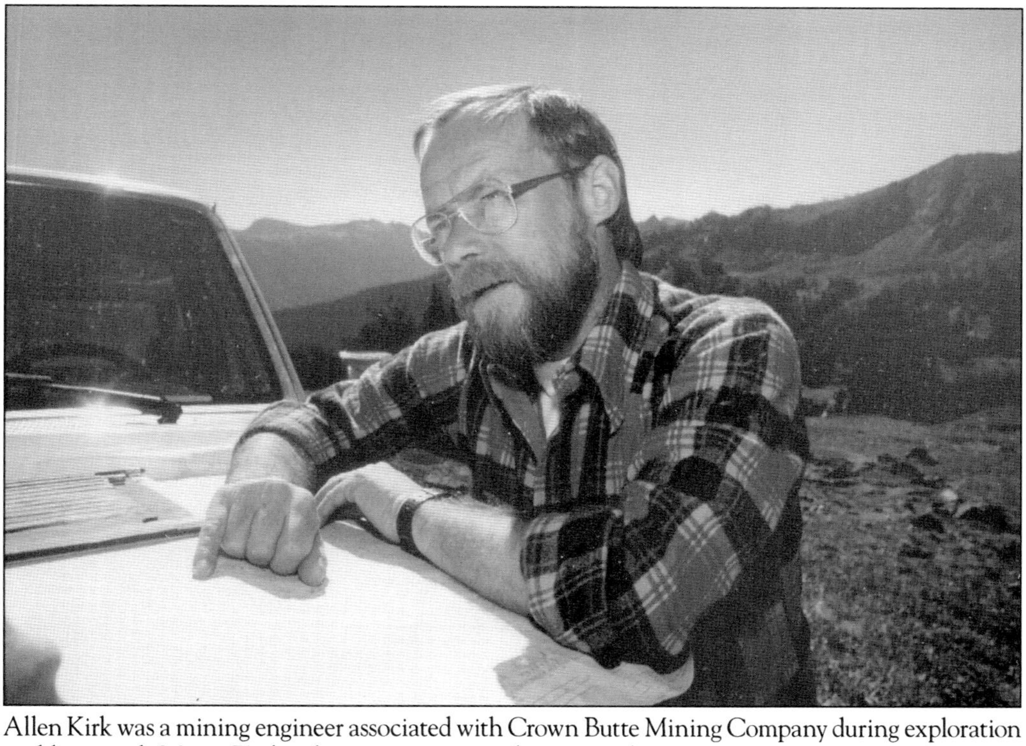
Allen Kirk was a mining engineer associated with Crown Butte Mining Company during exploration and later with Maxis Technology in mine remediation work.

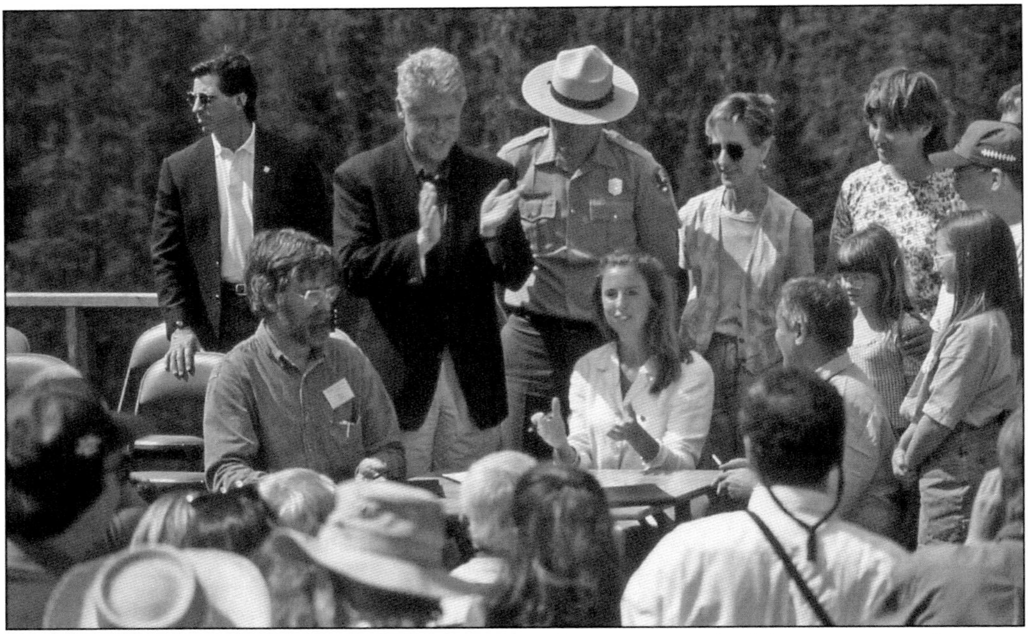

The fear that the proposed mine might jeopardize the crown jewel of the national park system brought action needed to reform the 1872 mining law. President Clinton announced the deal halting the controversial gold mine northeast of Yellowstone National Park in August 1996. (Courtesy of Yellowstone National Park, Jim Peaco.)

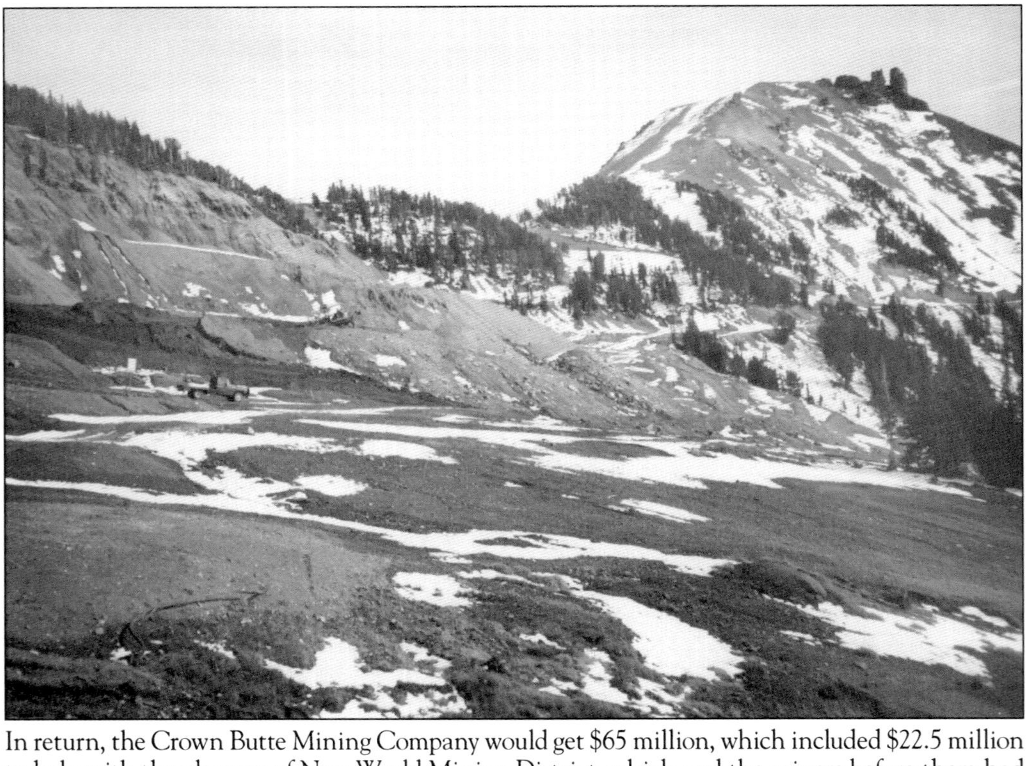

In return, the Crown Butte Mining Company would get $65 million, which included $22.5 million to help with the cleanup of New World Mining District, which and the miners before them had disturbed. Shown is reclamation work on the McLaren open pit.

Mine remediation work was done on mining sites in the Lulu and Fisher Creek area following the Crown Butte Mines exploration. The bed of Fisher Creek, shown in this photograph, was coated with iron leached out of the water from old mine adits.

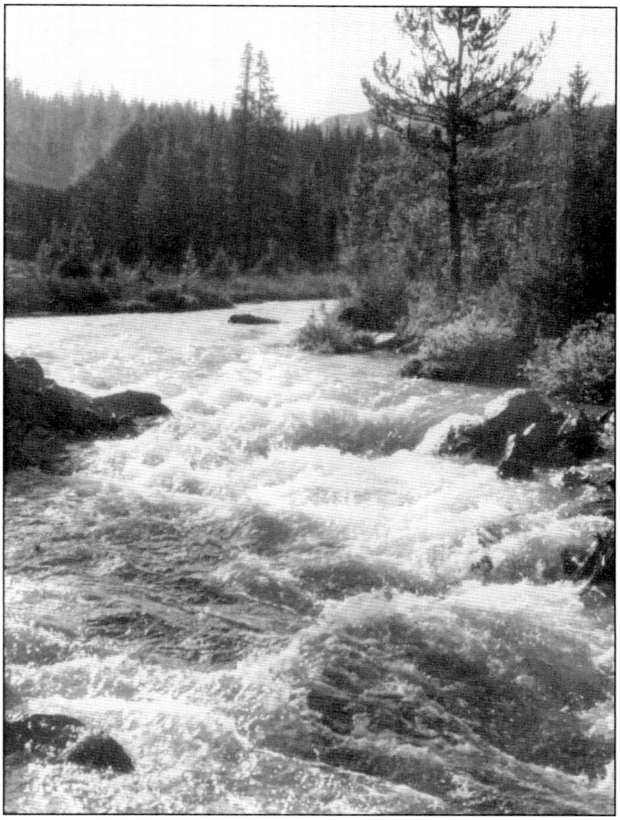

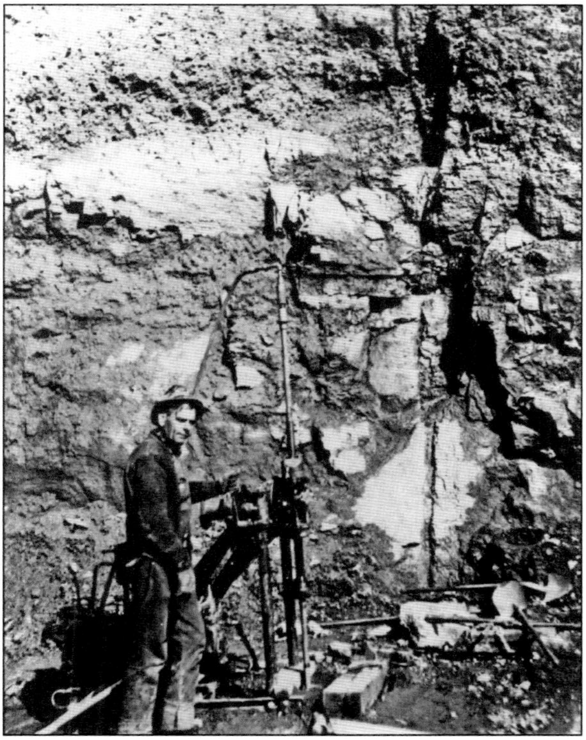

This undated photograph shows a miner drilling at the McLaren Mine, which operated from 1933 to 1953.

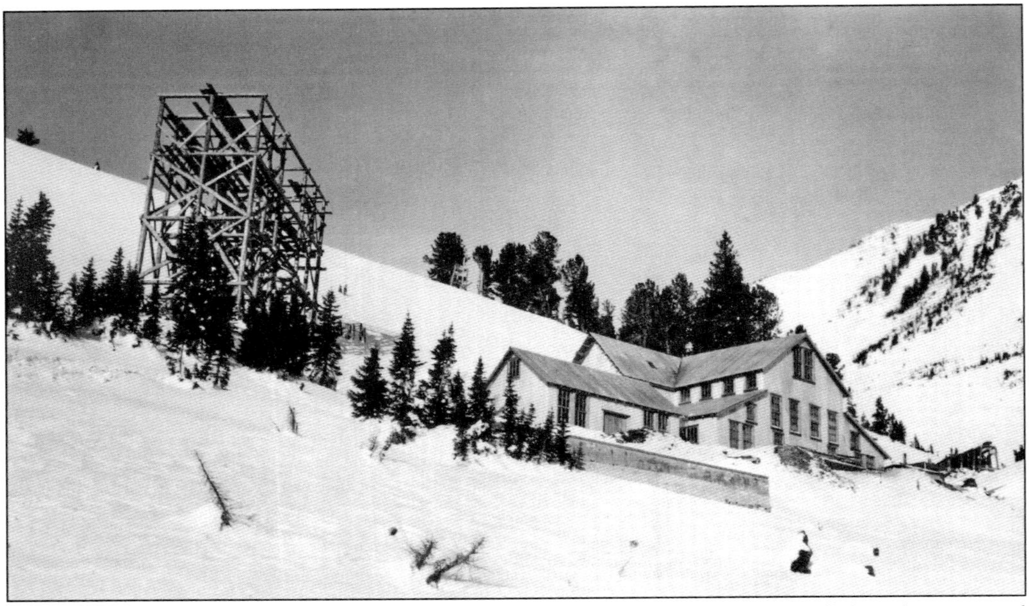

This winter photograph is identified as the Glengarry powerhouse, shop, mill foundations, and tramway. The Glengarry was southwest of Lulu Pass.

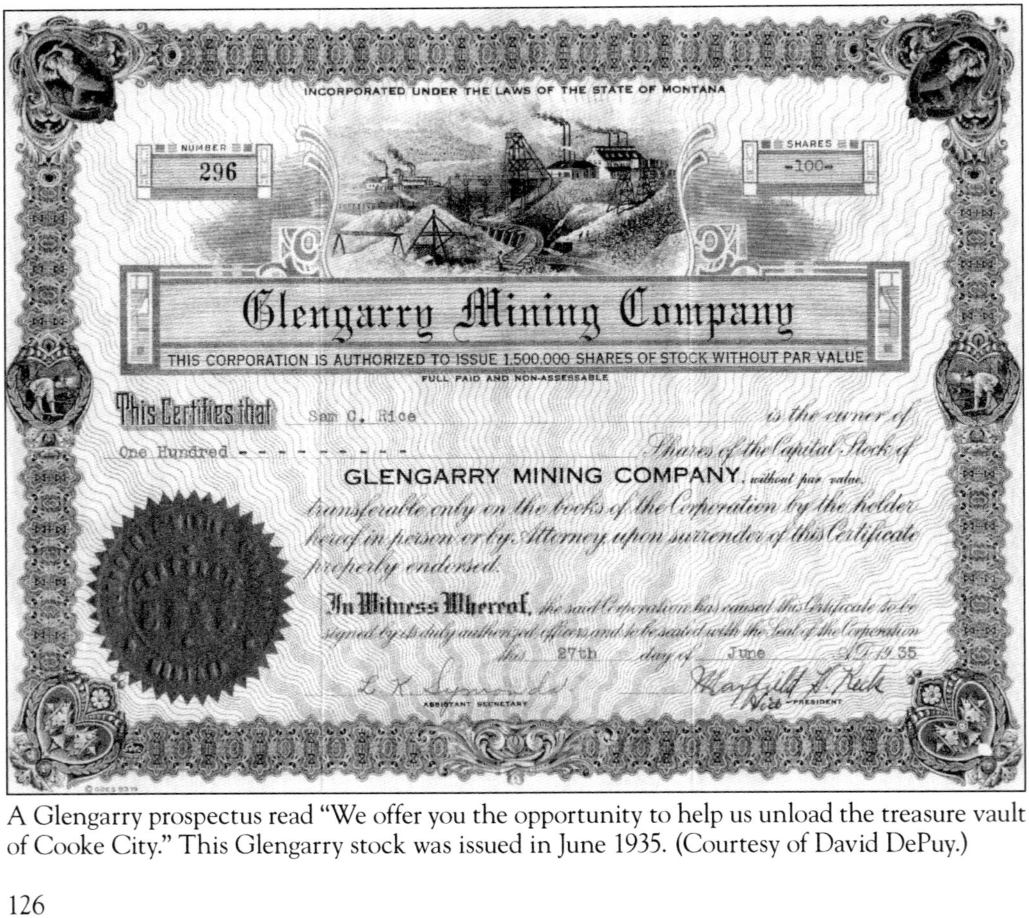

A Glengarry prospectus read "We offer you the opportunity to help us unload the treasure vault of Cooke City." This Glengarry stock was issued in June 1935. (Courtesy of David DePuy.)

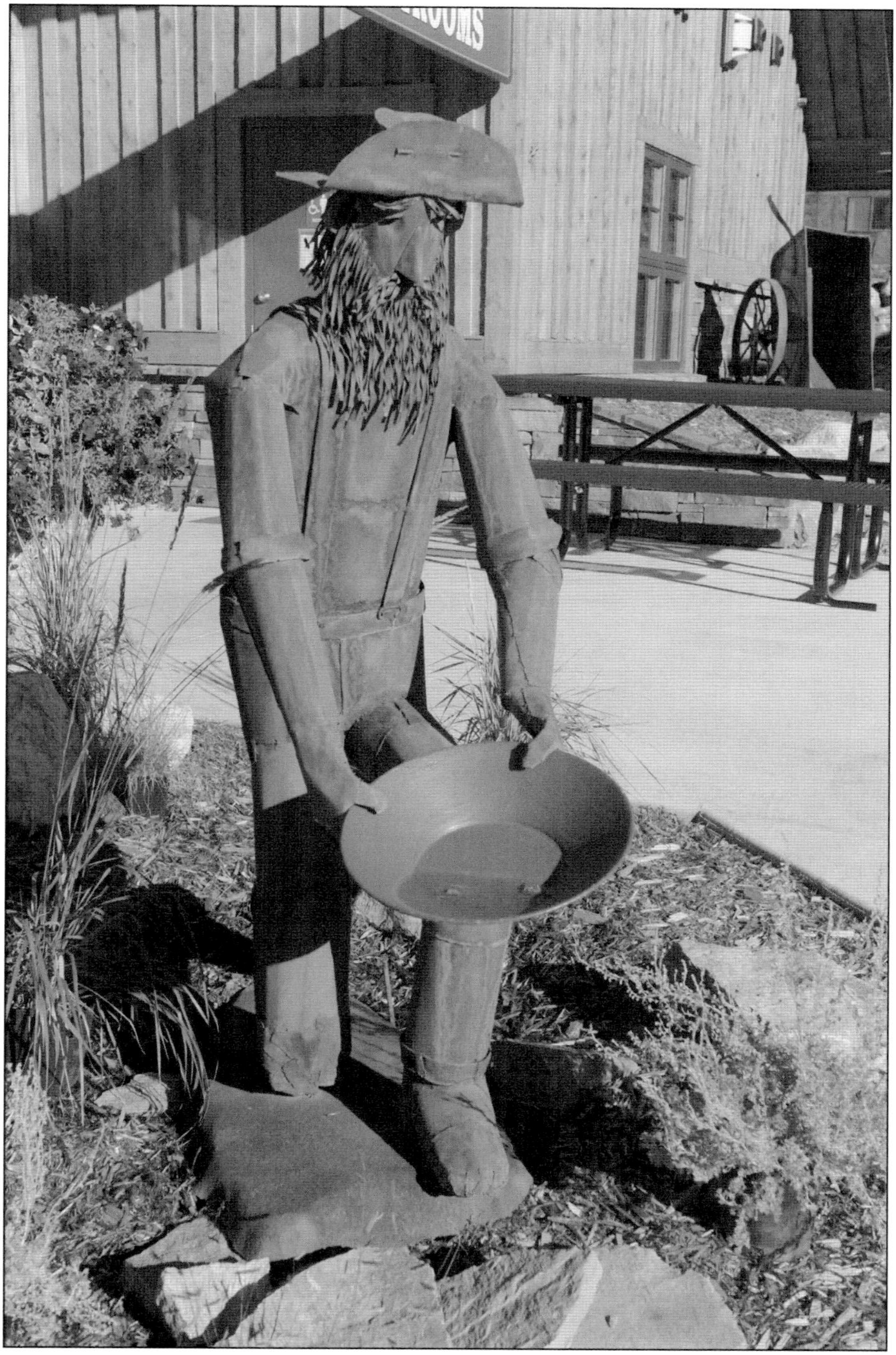

The Cooke City Visitor Center and the future Cooke City, Montana, Museum is in the process of organizing. This one-of-a-kind statue is titled *Panning the Dream*. The Cooke City, Montana, Museum is dedicated to the preservation and protection of objects and documents significant to the interpretation of the Cooke City, Colter Pass, and Silver Gate Townships. The collections of local history are dedicated for their inspirational, educational, and cultural values to this and future generations.

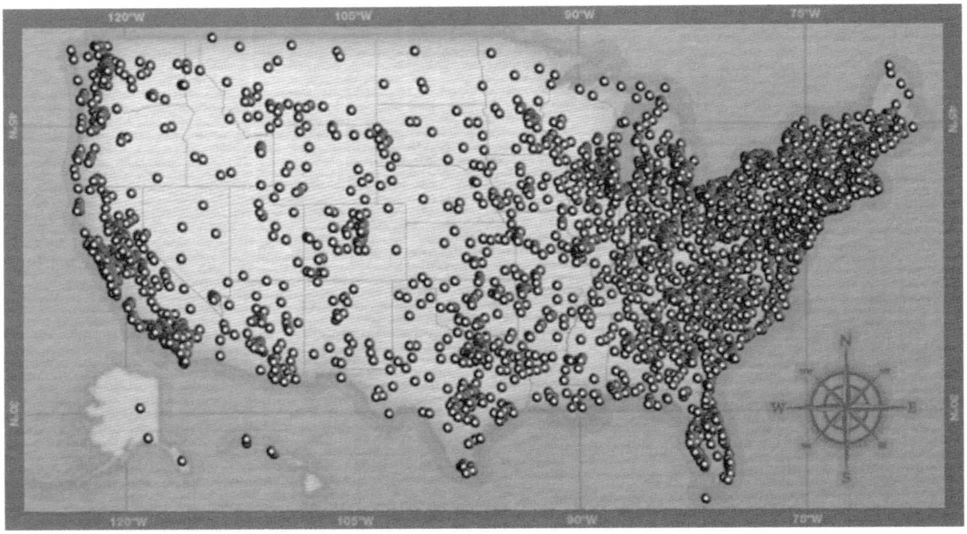